OLIVER BURKEMAN

Four Thousand Weeks

Time Management for Mortals

VINTAGE

9 10 8

Vintage is part of the Penguin Random House group of companies whose addresses can be found at global.penguinrandomhouse.com

First published in Vintage in 2022
First published in hardback by The Bodley Head in 2021

Some portions of this book first appeared in an earlier form in the *Guardian*.

penguin.co.uk/vintage

A CIP catalogue record for this book is available from the British Library

ISBN 9781784704001

Printed and bound in Great Britain by Clays Ltd, Elcograf S.p.A.

The authorised representative in the EEA is Penguin Random House Ireland, Morrison Chambers, 32 Nassau Street, Dublin D02 YH68

Penguin Random House is committed to a sustainable future for our business, our readers and our planet. This book is made from Forest Stewardship Council® certified paper.

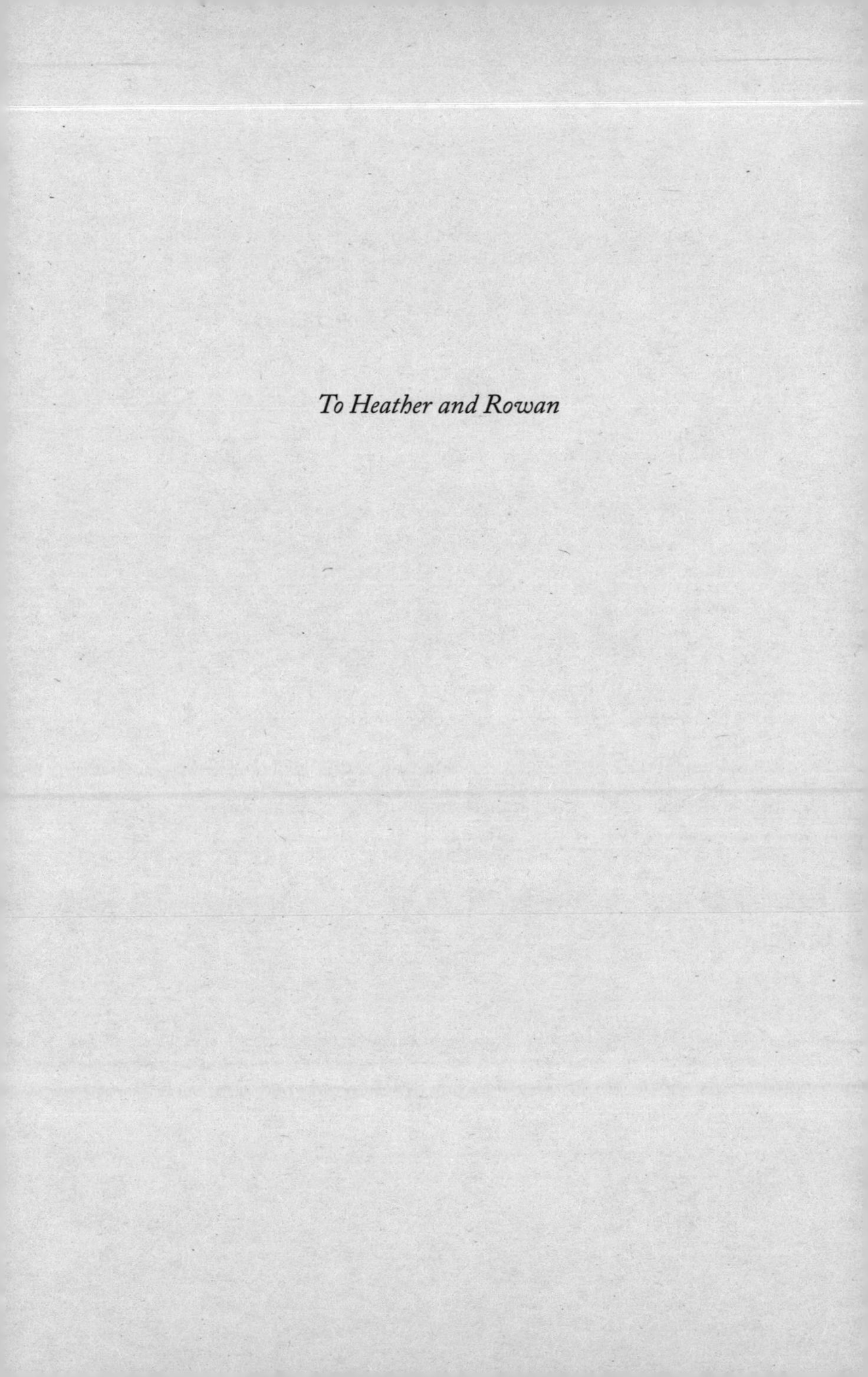

To Heather and Rowan

It's the very last thing, isn't it, we feel grateful for: having *happened*. You know, you needn't have happened. You needn't have happened. But you did happen.

– DOUGLAS HARDING

What makes it unbearable is your mistaken belief that it can be cured.

– CHARLOTTE JOKO BECK

Contents

Four Thousand Weeks

Introduction: In the Long Run, We're All Dead

The average human lifespan is absurdly, terrifyingly, insultingly short. Here's one way of putting things in perspective: the first modern humans appeared on the plains of Africa at least 200,000 years ago, and scientists estimate that life, in some form, will persist for another 1.5 billion years or more, until the intensifying heat of the sun condemns the last organism to death. But you? Assuming you live to be eighty, you'll have had about four thousand weeks.

Certainly, you might get lucky: make it to ninety, and you'll have had almost 4,700 weeks. You might get *really* lucky, like Jeanne Calment, the Frenchwoman who was thought to be 122 when she died in 1997, making her the oldest person on record. Calment claimed she could recall meeting Vincent van Gogh – she mainly remembered his reeking of alcohol – and she was still around for the birth of the first successfully cloned mammal, Dolly the sheep, in 1996. Biologists predict that lifespans within striking distance of Calment's could soon become commonplace. Yet even she got only about 6,400 weeks.

Expressing the matter in such startling terms makes it easy to see why philosophers from ancient Greece to the present day have taken the brevity of life to be the defining problem of human existence: we've been granted the mental capacities to make almost infinitely ambitious plans, yet practically no time at all to put them into action. 'This space that has been granted to us rushes by so speedily and so swiftly that all save a very few find life at an end just when they are getting ready to live,' lamented Seneca, the Roman philosopher, in a letter known today under the title *On the Shortness of Life*. When I first made the four thousand weeks calculation, I felt queasy; but once I'd recovered, I started pestering my friends, asking them to guess – off the top of their heads, without doing any mental arithmetic – how many weeks they thought the average person could expect to live. One named a number in the six figures. Yet, as I felt obliged to inform her, a fairly modest six-figure number of weeks – 310,000 – is the approximate duration of *all human civilisation* since the ancient Sumerians of Mesopotamia. On almost any meaningful timescale, as the contemporary philosopher Thomas Nagel has written, 'we will all be dead any minute'.

It follows from this that time management, broadly defined, should be everyone's chief concern. Arguably, time management is all life is. Yet the modern discipline known as time management – like its hipper cousin, productivity – is a depressingly narrow-minded affair, focused on how to crank through as many work tasks as possible, or on devising the perfect morning routine, or on cooking all your dinners for the week in one big batch on Sundays. These things matter to some extent, no doubt. But they're hardly all that matters. The world is bursting with wonder, and yet it's the rare productivity guru who seems to have considered the possibility that the ultimate point of all our frenetic *doing* might be to experience more of that wonder. The world also seems to be heading to hell in a handcart – our civic life has gone insane, a pandemic has paralysed society, and the planet is getting hotter and hotter – but good luck finding a time management system that makes any room for engaging productively with your fellow citizens, with current events, or with the fate of the environment. At the very least, you might have assumed there'd be a handful of productivity books that take seriously the stark facts about the shortness of life, instead of pretending that we can just ignore the subject. But you'd be wrong.

So this book is an attempt to help redress the balance – to see if we can't discover, or recover, some ways of thinking about time that do justice to our real situation: to the outrageous brevity and shimmering possibilities of our four thousand weeks.

Life on the Conveyor Belt

In one sense, of course, nobody these days needs telling that there isn't enough time. We're obsessed with our overfilled inboxes and lengthening to-do lists, haunted by the guilty feeling that we ought to be getting more done, or different things done, or both. (How can you be sure that people feel so busy? It's like the line about how to know whether someone's a vegan: don't worry, they'll tell you.) Surveys reliably show that we feel more pressed for time than ever before; yet in 2013, research by a team of Dutch academics raised the amusing possibility that such surveys may understate the scale of the busyness epidemic – because many people feel too busy to participate in surveys. Recently, as the gig economy has grown, busyness has been rebranded as 'hustle' – relentless work not as a burden to be endured but as an exhilarating lifestyle choice, worth boasting about on social media. In reality, though, it's the same old problem, pushed to an extreme: the pressure to fit ever-increasing quantities of activity into a stubbornly non-increasing quantity of daily time.

And yet busyness is really only the beginning. Many other complaints, when you stop to think about them, are essentially complaints about our limited time. Take the daily battle against online distraction, and the alarming sense that our attention spans have shrivelled to such a degree that even those of us who were bookworms as children now struggle to make it through a paragraph without experiencing the urge to reach for our phones. What makes this so troubling, in the end, is that it represents a failure to make the best use of a small supply of time. (You'd feel less self-loathing about wasting a morning on Facebook if the supply of mornings were inexhaustible.) Or perhaps your problem isn't being too busy but insufficiently busy, languishing in a dull job, or not employed at all. That's still a situation made far more distressing by the shortness of life, because you're *using up* your limited time in a way you'd rather not. Even some of the very worst aspects of our era – like our viciously hyperpartisan politics and terrorists radicalised via YouTube videos – can be explained, in a roundabout way, by the same underlying facts concerning life's brevity. It's because our time and attention are so limited, and therefore valuable, that social media companies are incentivised to grab as much of them as they can, by any means necessary – which is why they show users material guaranteed to drive them into a rage, instead of the more boring and accurate stuff.

Then there are all those timeless human dilemmas like whom to marry, whether to have children, and what kind of work to pursue. If we had thousands of years in which to live, all *those* would be far less agonising, too, since there'd be sufficient time to spend decades trying out each kind of possible existence. Meanwhile, no catalogue of our time-related troubles would be complete without mentioning that alarming phenomenon, familiar to anyone older than about thirty, whereby time seems to speed up as you age – steadily accelerating until, to judge from the reports of people in their seventies and eighties, months begin to flash by in what feels like minutes. It's hard to imagine a crueller arrangement: not only are our four thousand weeks constantly running out, but the fewer of them we have left, the faster we seem to lose them.

And if our relationship to our limited time has always been a difficult one, recent events have brought matters to a head. In 2020, in lockdown during the coronavirus pandemic, with our normal routines suspended, many people reported feeling that time was disintegrating completely, giving rise to the disorientating impression that their days were somehow simultaneously racing by and dragging on interminably. Time divided us, even more than it had before: for those with jobs and small children at home, there wasn't enough of it; for those furloughed or unemployed, there was too much. People found themselves working at strange hours, detached from the cycles of daytime and darkness, hunched over glowing laptops at home, or risking their lives in hospitals and mail-order warehouses. And it felt as though the future had been put on hold, leaving many of us stuck, in the words of one psychiatrist, 'in a new kind of everlasting present' – an anxious limbo of social media scrolling and desultory Zoom calls and insomnia, in which it felt impossible to make meaningful plans, or even to clearly picture life beyond the end of next week.

All of which makes it especially frustrating that so many of us are so *bad* at managing our limited time – that our efforts to make the most of it don't simply fail but regularly seem to make things worse. For years now, we've been deluged with advice on living the fully optimised life, in books with titles such as *Extreme Productivity* and *The 4-Hour Workweek* and *Smarter Faster Better*, plus websites full of 'life hacks' for whittling seconds off everyday chores. (Note the curious suggestion, in the term 'life hack', that your life is best thought of as some kind of faulty contraption, in need of modification so as to stop it from performing suboptimally.) There are numerous apps and wearable devices for maximising the pay-offs from your workday, your workouts and even your sleep, plus meal replacement drinks to eliminate time wasted eating dinner. And the chief selling point of a thousand other products and services, from kitchen appliances to online banking, is that they'll help you achieve the widely championed goal of squeezing the most from your time.

The problem isn't exactly that these techniques and products don't work. It's that they do work – in the sense that you'll get more done, race to more meetings, ferry your kids to more after-school activities, generate more profit for your employer – and yet, paradoxically, you only feel busier, more anxious, and somehow emptier as a result. In the modern world, the American anthropologist Edward T. Hall once pointed out, time feels like an unstoppable conveyor belt, bringing us new tasks as fast as we can dispatch the old ones; and becoming 'more productive' just seems to cause the belt to speed up. Or else, eventually, to break down: it's now common to encounter reports, especially from younger adults, of an all-encompassing, bone-deep burnout, characterised by an inability to complete basic daily chores – the paralysing exhaustion of 'a generation of finely honed tools, crafted from embryos to be lean, mean production machines', in the words of the millennial social critic Malcolm Harris.

This is the maddening truth about time, which most advice on managing it seems to miss. It's like an obstreperous toddler: the more you struggle to control it, to make it conform to your agenda, the further it slips from your control. Consider all the technology intended to help us gain the upper hand over time: by any sane logic, in a world with dishwashers, microwaves and jet engines, time ought to feel *more* expansive and abundant, thanks to all the hours freed up. But this is nobody's actual experience. Instead, life accelerates, and everyone grows more impatient. It's somehow vastly more aggravating to wait two minutes for the microwave than two hours for the oven – or ten seconds for a slow-loading web page versus three days to receive the same information by post.

The same self-defeating pattern applies to many of our attempts to become more productive at work. A few years ago, drowning in emails, I successfully implemented the system known as Inbox Zero, but I soon discovered that when you get tremendously efficient at answering emails all that happens is that you get much more email. Feeling busier – thanks to all that email – I bought *Getting Things Done*, by the time management guru David Allen, lured by his promise that it is 'possible for a person to have an overwhelming number of things to do and still function productively with a clear head' and 'what the martial artists call a "mind like water"'. But I failed to appreciate Allen's deeper implication – that there'll always be too much to do – and instead set about attempting to get an impossible amount done. In fact, I did get better at racing through my to-do list, only to find that greater volumes of work magically started to appear. (Actually, it's not magic; it's simple psychology, plus capitalism. More on that later.)

None of this is how the future was supposed to feel. In 1930, in a speech titled 'Economic Possibilities for Our Grandchildren', the economist John Maynard Keynes made a famous prediction: within a century, thanks to the growth of wealth and the advance of technology, no one would have to work more than about fifteen hours a week. The challenge would be how to fill all our new-found leisure time without going crazy. 'For the first time since his creation,' Keynes told his audience, 'man will be faced with his real, his permanent problem – how to use his freedom from pressing economic cares.' But Keynes was wrong. It turns out that when people make enough money to meet their needs, they just find new things to need and new lifestyles to aspire to; they never quite manage to keep up with the Joneses, because whenever they're in danger of getting close, they nominate new and better Joneses with whom to try to keep up. As a result, they work harder and harder, and soon busyness becomes an emblem of prestige. Which is clearly completely absurd: for almost the whole of history, the entire point of being rich was *not* having to work so much. Moreover, the busyness of the better-off is contagious, because one extremely effective way to make more money, for those at the top of the tree, is to cut costs and make efficiency improvements in their companies and industries. That means greater insecurity for those lower down, who are then obliged to work harder just to get by.

On Getting the Wrong Things Done

But now here we get to the heart of things, to a feeling that goes deeper, and that's harder to put into words: the sense that despite all this activity, even the relatively privileged among us rarely get round to doing the right things. We sense that there are important and fulfilling ways we could be spending our time, even if we can't say exactly what they are – yet we systematically spend our days doing other things instead. This yearning for more meaning can take many forms: it's there, for instance, in the desire to devote yourself to some larger cause, in the intuition that this particular moment in history, with all its crises and suffering, might demand more from you than the usual getting and spending. But it's also there in the feeling of frustration at having to work a day job in order to buy slivers of time for the work you love, and in the simple longing to spend more of your brief time on earth with your kids, in nature, or, at the very least, not commuting. The environmentalist and spiritual writer Charles Eisenstein recalls first sensing this basic 'wrongness' in our use of time as a child, growing up amid material comfort in 1970s America:

> Life, I knew, was supposed to be more joyful than this, more real, more meaningful, and the world was supposed to be more beautiful. We were not supposed to hate Mondays and live for the weekends and holidays. We were not supposed to have to raise our hands to be allowed to pee. We were not supposed to be kept indoors on a beautiful day, day after day.

And this feeling of wrongness is only exacerbated by our attempts to become more productive, which seem to have the effect of pushing the genuinely important stuff ever further over the horizon. Our days are spent trying to 'get through' tasks, in order to get them 'out of the way', with the result that we live mentally in the future, waiting for when we'll finally get round to what really matters – and worrying, in the meantime, that we don't measure up, that we might lack the drive or stamina to keep pace with the speed at which life now seems to move. 'The spirit of the times is one of joyless urgency,' writes the essayist Marilynne Robinson, who observes that many of us spend our lives 'preparing ourselves and our children to be means to inscrutable ends that are utterly not our own'. Our struggle to stay on top of everything may serve *someone's* interests; working longer hours – and using any extra income to buy more consumer goods – turns us into better cogs in the economic machine. But it doesn't result in peace of mind, or lead us to spend more of our finite time on those people and things we care most deeply about ourselves.

Four Thousand Weeks is yet another book about making the best use of time. But it is written in the belief that time management as we know it has failed miserably, and that we need to stop pretending otherwise. This strange moment in history, when time feels so unmoored, might in fact provide the ideal opportunity to reconsider our relationship with it. Older thinkers have faced these challenges before us, and when their wisdom is applied to the present day, certain truths grow more clearly apparent. Productivity is a trap. Becoming more efficient just makes you more rushed, and trying to clear the decks simply makes them fill up again faster. Nobody in the history of humanity has ever achieved 'work–life balance', whatever that might be, and you certainly won't get there by copying the 'six things successful people do before 7 a.m'. The day will never arrive when you finally have everything under control – when the flood of emails has been contained; when your to-do lists have stopped getting longer; when you're meeting all your obligations at work and in your home life; when nobody's angry with you for missing a deadline or dropping the ball; and when the fully optimised person you've become can turn, at long last, to the things life is really supposed to be about. Let's start by admitting defeat: none of this is ever going to happen.

But you know what? That's *excellent* news.

Part I

Choosing to Choose

1.

The Limit-Embracing Life

The real problem isn't our limited time. The real problem – or so I hope to convince you – is that we've unwittingly inherited, and feel pressured to live by, a troublesome set of *ideas* about how to use our limited time, all of which are pretty much guaranteed to make things worse. To see how we got here, and how to escape into a better relationship with time, we need to rewind the clock – back to before there were clocks.

On balance, you should definitely be grateful you weren't born a peasant in early-medieval England. For one thing, you'd have been much less likely to make it to adulthood; but even if you had, the life that stretched ahead of you would have been one defined by servitude. You'd have spent your back-breaking days farming the land on which the local lord permitted you to live, in exchange for giving him a crippling proportion of what you produced or the income you could generate from it. The church would have demanded regular contributions as well, and you'd have been much too scared of eternal damnation to disobey. At night, you would have retreated to your one-room hut, alongside not only the rest of your family (who, like you, would rarely have bathed or brushed their teeth) but also your pigs and chickens, which you brought indoors at night; bears and wolves still roamed the forests and would help themselves to any animals left outside after sunset. Disease would have been another constant companion: familiar sicknesses ranged from measles and influenza to bubonic plague and St Anthony's fire, a form of food poisoning caused by mouldy grain, which left the delirious sufferer feeling as though his skin were burning or as if he were being bitten by unseen teeth.

Time Before Timetables

But there's one set of problems you almost certainly wouldn't have experienced: problems of time. Even on your most exhausting days, it probably wouldn't have occurred to you that you had 'too much to do', that you needed to hurry, or that life was moving too fast, let alone that you'd got your work–life balance wrong. By the same token, on quieter days, you would never have felt bored. And though death was a constant presence, with lives cut short far more frequently than they are today, time wouldn't have felt in limited supply. You wouldn't have felt any pressure to find ways to 'save' it. Nor would you have felt guilty for wasting it: if you took an afternoon break from threshing grain to watch a cockfight on the village green, it wouldn't have felt like you were shirking during 'work time'. And none of this was simply because things moved more slowly back then, or because medieval peasants were more relaxed or more resigned to their fate. It was because, so far as we can tell, they generally didn't experience time as an abstract entity – as a *thing* – at all.

If that sounds confusing, it's because our modern way of thinking about time is so deeply entrenched that we forget it even is a way of thinking; we're like the proverbial fish who have no idea what water is, because it surrounds them completely. Get a little mental distance on it, though, and our perspective starts to look rather peculiar. We imagine time to be something separate from us and from the world around us, 'an independent world of mathematically measurable sequences', in the words of the American cultural critic Lewis Mumford. To see what he means, consider some time-related question – how you plan to spend tomorrow afternoon, say, or what you've accomplished over the last year – and without being fully conscious of it at first, you'll probably find yourself visualising a calendar, a yardstick, a tape measure, the numbers on a clock face, or some hazier kind of abstract timeline. You'll then proceed to measure and judge your real life against this imaginary gauge, lining up your activities against the timeline in your head. Edward Hall was making the same point with his image of time as a conveyor belt that's constantly passing us by. Each hour or week or year is like a container being carried on the belt, which we must fill as it passes, if we're to feel that we're making good use of our time. When there are too many activities to fit comfortably into the containers, we feel unpleasantly busy; when there are too few, we feel bored. If we keep pace with the passing containers, we congratulate ourselves for 'staying on top of things' and feel like we're justifying our existence; if we let too many pass by unfilled, we feel we've wasted them. If we use containers labelled 'work time' for the purposes of leisure, our employer may grow irritated. (He paid for those containers; they belong to him!)

The medieval farmer simply had no reason to adopt such a bizarre idea in the first place. Workers got up with the sun and slept at dusk, the lengths of their days varying with the seasons. There was no need to think of time as something abstract and separate from life: you milked the cows when they needed milking and harvested the crops when it was harvest time, and anybody who tried to impose an external schedule on any of that – for example, by doing a month's milking in a single day to get it out of the way, or by trying to make the harvest come sooner – would rightly have been considered a lunatic. There was no anxious pressure to 'get everything done', either, because a farmer's work is infinite: there will always be another milking and another harvest, forever, so there's no sense in racing towards some hypothetical moment of completion. Historians call this way of living 'task orientation', because the rhythms of life emerge organically from the tasks themselves, rather than from being lined up against an abstract timeline, the approach that has become second nature for us today. (It's tempting to think of medieval life as moving slowly, but it's more accurate to say that the concept of life 'moving slowly' would have struck most people as meaningless. Slowly as compared with what?) In those days before clocks, when you did need to explain how long something might take, your only option was to compare it with some other concrete activity. Medieval people might speak of a task lasting a 'Miserere whyle' – the approximate time it took to recite Psalm 50, known as the Miserere, from the Bible – or alternatively a 'pissing whyle', which should require no explanation.

Living this way, one can imagine that experience would have felt expansive and fluid, suffused with something it might not be an exaggeration to call a kind of magic. Notwithstanding the many real privations of his existence, our peasant farmer might have sensed a luminous, awe-inspiring dimension to the world around him. Untroubled by any notion of time 'ticking away', he might have experienced a heightened awareness of the vividness of things, the feeling of timelessness that Richard Rohr, a contemporary Franciscan priest and author, calls 'living in deep time'. At dusk, the medieval country-dweller might have sensed spirits whispering in the forest, along with the bears and wolves; ploughing the fields, he might have felt himself one tiny part of a vast sweep of history, in which his distant ancestors were almost as alive to him as his own children. We can assert all this with some confidence because we still occasionally encounter islands of deep time today – in those moments when, to quote the writer Gary Eberle, we slip 'into a realm where there is enough of everything, where we are not trying to fill a void in ourselves or the world'. The boundary separating the self from the rest of reality grows blurry, and time stands still. 'The clock does not stop, of course,' Eberle writes, 'but we do not hear it ticking.'

This happens for some people in prayer, or in meditation,

or in magnificent landscapes; I'm pretty sure my toddler son spent the whole of his infancy in such a state and is only now beginning to leave it. (Until we get them onto schedules, babies are the ultimate 'task-orientated' beings, which, along with sleep deprivation, may explain the other-worldliness of those first few months with a newborn: you're dragged from clock time into deep time, whether you like it or not.) The Swiss psychologist Carl Jung, visiting Kenya in 1925, was setting out on a hike in the first glow of dawn when he, too, was suddenly plunged into timelessness:

> From a low hill in this broad savanna, a magnificent prospect opened out to us. To the very brink of the horizon we saw gigantic herds of animals: gazelle, antelope, gnu, zebra, warthog, and so on. Grazing heads nodding, the herds moved forward like slow rivers. There was scarcely any sound save the melancholy cry of a bird of prey. This was the stillness of the eternal beginning, the world as it had always been, in the state of non-being . . . I walked away from my companions until I had put them out of sight, and savoured the feeling of being entirely alone.

The End of Eternity

There's one huge drawback in giving so little thought to the abstract idea of time, though, which is that it severely limits what you can accomplish. You can be a small-scale farmer, relying on the seasons for your schedule, but you can't be much other than a small-scale farmer (or a baby). As soon as you want to coordinate the actions of more than a handful of people, you need a reliable, agreed-upon method of measuring time. This is widely held to be how the first mechanical clocks came to be invented, by medieval monks, who had to begin their morning prayers while it was still dark, and needed some way of ensuring that the whole monastery woke up at the required point. (Their earlier strategies included deputising one monk to stay awake all night, watching the movements of the stars – a system that worked only when it wasn't cloudy, and the night-shift monk didn't fall asleep.) Making time standardised and visible in this fashion inevitably encourages people to think of it as an abstract thing with an independent existence, distinct from the specific activities on which one might spend it; 'time' is what ticks away as the hands move round the clock face. The Industrial Revolution is usually attributed to the invention of the steam engine; but as Mumford shows in his 1934 magnum opus, *Technics and Civilization*, it also probably couldn't have happened without the clock. By the late 1700s, rural peasants were streaming into English cities, taking jobs in mills and factories, each of which required the coordination of hundreds of people, working fixed hours, often six days a week, to keep the machines running.

From thinking about time in the abstract, it's natural to start treating it as a *resource*, something to be bought and sold and used as efficiently as possible, like coal or iron or any other raw material. Previously, labourers had been paid for a vaguely defined 'day's work', or on a piecework basis, receiving a given sum per bale of hay or per slaughtered pig. But gradually it became more common to be paid by the hour – and the factory owner who used his workers' hours efficiently, squeezing as much labour as possible from each employee, stood to make a bigger profit than one who didn't. Indeed, some cantankerous industrialists came to feel that workers who didn't drive themselves hard enough were literally guilty of stealing something. 'I have by sundry people [been] horribly cheated,' fumed the iron magnate Ambrose Crowley, from County Durham, in a memo from the 1790s, announcing his new policy of deducting pay for time spent 'smoking, singing, reading of news history, contention, disputes, anything foreign to my business [or] in any way loitering'. The way Crowley saw it, his lackadaisical employees were thieves, illegitimately helping themselves to containers from the conveyor belt of time.

You don't need to believe, as Mumford sometimes seems to imply, that the invention of the clock is solely to blame for all our time-related troubles today. (And I certainly won't be arguing for a return to the lifestyle of medieval peasants.) But a threshold had been crossed. Before, time was just the medium in which life unfolded, the stuff that life was made of. Afterwards, once 'time' and 'life' had been separated in most people's minds, time became a *thing* that you *used* – and it's this shift that serves as the precondition for all the uniquely modern ways in which we struggle with time today. Once time is a resource to be used, you start to feel pressure, whether from external forces or from yourself, to use it well, and to berate yourself when you feel you've wasted it. When you're faced with too many demands, it's easy to assume that the only answer must be to make *better use* of time, by becoming more efficient, driving yourself harder, or working for longer – as if you were a machine in the Industrial Revolution – instead of asking whether the demands themselves might be unreasonable. It grows alluring to try to multitask – that is, to use the same portion of time for two things at once, as the German philosopher Friedrich Nietzsche was one of the first to notice: 'One thinks with a watch in one's hand,' he complained in an 1887 essay, 'even as one eats one's midday meal while reading the latest news of the stock market.' And it becomes a lot more intuitive to project your thoughts about your life into an imagined future, leaving you anxiously wondering if things will unfold as you want them to. Soon, your sense of self-worth gets completely bound up with how you're using time: it stops being merely the water in which you swim and turns into something you feel you need to dominate or control, if you're to avoid feeling guilty, panicked or overwhelmed. The title of a book that arrived on my desk the other day sums things up nicely: *Master Your Time, Master Your Life*.

The fundamental problem is that this attitude towards time sets up a rigged game in which it's impossible ever to feel as though you're doing well enough. Instead of simply living our lives as they unfold in time – instead of just *being* time, you might say – it becomes difficult not to value each moment primarily according to its usefulness for some future goal, or for some future oasis of relaxation you hope to reach once your tasks are finally 'out of the way'. Superficially, this seems like a sensible way to live, especially in a hypercompetitive economic climate, in which it feels as though you must constantly make the most judicious use of your time if you want to stay afloat. (It also reflects the manner in which most of us were brought up: to prioritise future benefits over current enjoyments.) But ultimately it backfires. It wrenches us out of the present, leading to a life spent leaning into the future, worrying about whether things will work out, experiencing everything in terms of some later, hoped-for benefit, so that peace of mind never quite arrives. And it makes it all but impossible to experience 'deep time', that sense of timeless time which depends on forgetting the abstract yardstick and plunging back into the vividness of reality instead.

As this modern mindset came to dominate, wrote Mumford, 'Eternity ceased gradually to serve as the measure and focus of human actions.' In its place came the dictatorship of the clock, the schedule and the Google Calendar alert; Marilynne Robinson's 'joyless urgency' and the constant feeling that you ought to be getting more done. The trouble with attempting to master your time, it turns out, is that time ends up mastering you.

Confessions of a Productivity Geek

The rest of this book is an exploration of a saner way of relating to time and a toolbox of practical ideas for doing so, drawn from the work of philosophers, psychologists and spiritual teachers who all rejected the struggle to dominate or master it. I believe it sketches a kind of life that's vastly more peaceful and meaningful – while also, it turns out, being better for sustained productivity over the long haul. But don't get me wrong: I spent years trying, and failing, to achieve mastery over my time. In fact, the symptoms were especially glaring in the subspecies to which I belonged. I was a 'productivity geek'. You know how some people are passionate about bodybuilding, or fashion, or rock climbing, or poetry? Productivity geeks are passionate about crossing items off their to-do lists. So it's sort of the same, except infinitely sadder.

My adventures with Inbox Zero were only the tip of the iceberg. I've squandered countless hours – and a fair amount of money, spent mainly on fancy notebooks and felt-tip pens – in service to the belief that if I could only find the right time management system, build the right habits, and apply sufficient self-discipline, I might actually be able to win the struggle with time, once and for all. (I was enabled in this delusion by writing a weekly newspaper column on productivity, which gave me an excuse to experiment with new techniques on the grounds that I was doing so for work purposes; I was like an alcoholic conveniently employed as a wine expert.) On one occasion, I tried scheduling the whole of every day in fifteen-minute blocks; on another, I used a kitchen timer to work exclusively in periods of twenty-five minutes, interspersed with five-minute breaks. (This approach has an official name, the Pomodoro Technique, and a cult following online.) I divided my lists into A, B and C priorities. (Guess how many B- and C-priority tasks I ever got around to completing?) I tried to align my daily actions with my goals, and my goals with my core values. Using these techniques often made me feel as if I were on the verge of ushering in a golden era of calm, undistracted productivity and meaningful activity. But it never arrived. Instead, I just got more stressed and unhappy.

I remember sitting on a park bench near my home in Brooklyn one winter morning in 2014, feeling even more anxious than usual about the volume of undone tasks, and suddenly realising that *none of this was ever going to work.* I would never succeed in marshalling enough efficiency, self-discipline and effort to force my way through to the feeling that I was on top of everything, that I was fulfilling all my obligations and had no need to worry about the future. Ironically, the realisation that this had been a useless strategy for attaining peace of mind brought me some immediate peace of mind. (After all, once you become convinced that something you've been attempting is impossible, it's a lot harder to keep on berating yourself for failing.) What I had yet to understand, at that point, was *why* all these methods were doomed to fail, which was that I was using them to try to obtain a feeling of control over my life that would always remain out of reach.

Though I'd been largely unaware of it, my productivity obsession had been serving a hidden emotional agenda. For one thing, it helped me combat the sense of precariousness inherent to the modern world of work: if I could meet every editor's every demand, while launching various side projects of my own, maybe one day I'd finally feel secure in my career and my finances. But it also held at bay certain scary questions about what I was doing with my life, and whether major changes might not be needed. If I could get enough work done, my subconscious had apparently concluded, I wouldn't need to ask if it was all that healthy to be deriving so much of my sense of self-worth from work in the first place. And as long as I was always just on the cusp of mastering my time, I could avoid the thought that what life was really demanding from me might involve *surrendering* the craving for mastery and diving into the unknown instead. In my case, that turned out to mean committing to a long-term relationship and, later, making the decision with my wife to try to start a family – two things I'd notably failed to get done with any number of systems for getting things done. It had been more comforting to imagine that I might eventually 'optimise' myself into the kind of person who could confront such decisions without fear, feeling totally in charge of the process. I didn't want to accept that this was never going to happen – that fear was part of the deal, and that experiencing it wouldn't destroy me.

But (don't worry!) we won't be dwelling here on my personal hang-ups. The universal truth behind my specific issues is that most of us invest a lot of energy, one way or another, in trying to avoid fully experiencing the reality in which we find ourselves. We don't want to feel the anxiety that might arise if we were to ask ourselves whether we're on the right path, or what ideas about ourselves it could be time to give up. We don't want to risk getting hurt in relationships or failing professionally; we don't want to accept that we might never succeed in pleasing our parents or in changing certain things we don't like about ourselves – and we certainly don't want to get ill and die. The details differ from person to person, but the kernel is the same. We recoil from the notion that this is it – that *this* life, with all its flaws and inescapable vulnerabilities, its extreme brevity, and our limited influence over how it unfolds, is the only one we'll get a shot at. Instead, we mentally fight against the way things are – so that, in the words of the psychotherapist Bruce Tift, 'we don't have to consciously participate in what it's like to feel claustrophobic, imprisoned, powerless, and constrained by reality'. This struggle against the distressing constraints of reality is what some old-school psychoanalysts call 'neurosis', and it takes countless forms, from workaholism and commitment-phobia to co-dependency and chronic shyness.

Our troubled relationship with time arises largely from this same effort to avoid the painful constraints of reality. And most of our strategies for becoming more productive make things worse, because they're really just ways of furthering the avoidance. After all, it's painful to confront how limited your time is, because it means that tough choices are inevitable and that you won't have time for all you once dreamed you might do. It's also painful to accept your limited control over the time you do get: maybe you simply lack the stamina or talent or other resources to perform well in all the roles you feel you should. And so, rather than face our limitations, we engage in avoidance strategies, in an effort to carry on feeling limitless. We push ourselves harder, chasing fantasies of the perfect work–life balance; or we implement time management systems that promise to make time for everything, so that tough choices won't be required. Or we procrastinate, which is another means of maintaining the feeling of omnipotent control over life – because you needn't risk the upsetting experience of failing at an intimidating project, obviously, if you never even start it. We fill our minds with busyness and distraction to numb ourselves emotionally. ('We labour at our daily work more ardently and thoughtlessly than is necessary to sustain our life,' wrote Nietzsche, 'because to us it is even more necessary not to have leisure to stop and think. Haste is universal because everyone is in flight from himself.') Or we plan compulsively, because the alternative is to confront how little control over the future we really have. Moreover, most of us seek a specifically individualistic kind of mastery over time – our culture's ideal is that you alone should control your schedule, doing whatever you prefer, whenever you want – because it's scary to confront the truth that almost everything worth doing, from marriage and parenting to business or politics, depends on cooperating with others, and therefore on exposing yourself to the emotional uncertainties of relationships.

Denying reality never works, though. It may provide some immediate relief, because it allows you to go on thinking that at some point in the future you might, at last, feel totally in control. But it can't ever bring the sense that you're doing enough – that you *are* enough – because it defines 'enough' as a kind of limitless control that no human can attain. Instead, the endless struggle leads to more anxiety and a less fulfilling life. For example, the more you believe you might succeed in 'fitting everything in', the more commitments you naturally take on, and the less you feel the need to ask whether each new commitment is truly worth a portion of your time – and so your days inevitably fill with more activities you don't especially value. The more you hurry, the more frustrating it is to encounter tasks (or toddlers) that won't be hurried; the more compulsively you plan for the future, the more anxious you feel about any remaining uncertainties, of which there will always be plenty. And the more individual sovereignty you achieve over your time, the lonelier you get.

All of this illustrates what might be termed the *paradox of limitation*, which runs through everything that follows: the more you try to manage your time with the goal of achieving a feeling of total control, and freedom from the inevitable constraints of being human, the more stressful, empty and frustrating life gets. But the more you confront the facts of finitude instead – and work with them, rather than against them – the more productive, meaningful and joyful life becomes. I don't think the feeling of anxiety ever completely goes away; we're even limited, apparently, in our capacity to embrace our limitations. But I'm aware of no other time management technique that's half as effective as just facing the way things truly are.

An Icy Blast of Reality

In practical terms, a limit-embracing attitude to time means organising your days with the understanding that you definitely *won't* have time for everything you want to do, or that other people want you to do – and so, at the very least, you can stop beating yourself up for failing. Since hard choices are unavoidable, what matters is learning to make them consciously, deciding what to focus on and what to neglect, rather than letting them get made by default – or deceiving yourself that, with enough hard work and the right time management tricks, you might not have to make them at all. It also means resisting the seductive temptation to 'keep your options open' – which is really just another way of trying to feel in control – in favour of deliberately making big, daunting, irreversible commitments, which you can't know in advance will turn out for the best, but which reliably prove more fulfilling in the end. And it means standing firm in the face of FOMO, the 'fear of missing out', because you come to realise that missing out on something – indeed, on almost everything – is basically guaranteed. Which isn't actually a problem anyway, it turns out, because 'missing out' is what makes our choices meaningful in the first place. Every decision to use a portion of time on anything represents the sacrifice of all the other ways in which you could have spent that time, but didn't – and to willingly make that sacrifice is to take a stand, without reservation, on what matters most to you. I should probably clarify that I have yet to attain perfection in any of these attitudes; I wrote this book for myself, as much as for anyone else, putting my faith in the words of the author Richard Bach: 'You teach best what you most need to learn.'

This confrontation with limitation also reveals the truth that freedom, sometimes, is to be found not in achieving greater sovereignty over your own schedule but in allowing yourself to be constrained by the rhythms of community – participating in forms of social life where you *don't* get to decide exactly what you do or when you do it. And it leads to the insight that meaningful productivity often comes not from hurrying things up but from letting them take the time they take, surrendering to what in German has been called *Eigenzeit*, or the time inherent to a process itself. Perhaps most radically of all, seeing and accepting our limited powers over our time can prompt us to question the very idea that time is something you *use* in the first place. There is an alternative: the unfashionable but powerful notion of *letting time use you*, approaching life not as an opportunity to implement your predetermined plans for success but as a matter of responding to the needs of your place and your moment in history.

I want to be clear that I'm not suggesting our troubles with time are somehow all in the mind, or that a simple change of outlook will cause them all to vanish. Time pressure comes largely from forces outside ourselves: from a cutthroat economy; from the loss of the social safety nets and family networks that used to help ease the burdens of work and childcare; and from the sexist expectation that women must excel in their careers while assuming most of the responsibilities at home. None of that will be solved by self-help alone; as the journalist Anne Helen Petersen writes in a widely shared essay on millennial burnout, you can't fix such problems 'with vacation, or an adult coloring book, or "anxiety baking", or the Pomodoro Technique, or overnight fucking oats'. But my point here is that however privileged or unfortunate your specific situation, fully facing the reality of it can only help. So long as you continue to respond to impossible demands on your time by trying to persuade yourself that you might one day find some way to do the impossible, you're implicitly collaborating with those demands. Whereas once you deeply grasp that they *are* impossible, you'll be newly empowered to resist them, and to focus instead on building the most meaningful life you can, in whatever situation you're in.

This notion that fulfilment might lie in embracing, rather than denying, our temporal limitations wouldn't have surprised the philosophers of ancient Greece and Rome. They understood limitlessness to be the sole preserve of the gods; the noblest of human goals wasn't to become godlike, but to be wholeheartedly human instead. In any case, this is just how reality is, and it can be surprisingly energising to confront it. Back in the 1950s, a splendidly cranky British author named Charles Garfield Lott Du Cann wrote a short book, *Teach Yourself to Live*, in which he recommended the limit-embracing life, and he responded saltily to the suggestion that his advice was depressing. 'Depressing? Not a bit of it. No more depressing than a cold [shower] is depressing . . . You are no longer befogged and bewildered by a false and misleading illusion about your life – like most people.' This is an excellent spirit in which to confront the challenge of using time well. None of us can single-handedly overthrow a society dedicated to limitless productivity, distraction and speed. But right here, right now, you can stop buying into the delusion that any of that is ever going to bring satisfaction. You can face the facts. You can turn on the shower, brace yourself for some invigoratingly icy water, and step in.

2.

The Efficiency Trap

Let's begin with busyness. It isn't our only time problem, and it isn't everyone's problem. But it's a uniquely vivid illustration of the effort we invest in fighting against our built-in limitations, thanks to how normal it has become to feel as though you absolutely *must* do more than you *can* do. 'Busyness' is a misnomer for this state of affairs, really, because certain forms of busyness can be delightful. Who wouldn't want to live in Busytown, the setting of the iconic 1960s children's books by the American illustrator Richard Scarry? His grocer cats and firefighting pigs are certainly busy; nobody in Busytown is idle – or if they are, they're carefully hidden from view by the authorities, Pyongyang-style. What they aren't, though, is *overwhelmed*. They exude the cheery self-possession of cats and pigs who have plenty to do, but also every confidence that their tasks will fit snugly into the hours available – whereas we live with the constant anxiety of fearing, or knowing for certain, that ours won't.

Research shows that this feeling arises on every rung of the economic ladder. If you're working two minimum-wage jobs to put food in your children's stomachs, there's a good chance you'll feel overstretched. But if you're better off, you'll find yourself feeling overstretched for reasons that seem, to you, no less compelling: because you have a nicer house with higher mortgage payments, or because the demands of your (interesting, well-paid) job conflict with your longing to spend time with your ageing parents, or to be more involved in your children's lives, or to dedicate your life to fighting climate change. As the Yale University law professor Daniel Markovits has shown, even the winners in our achievement-obsessed culture – the ones who make it to the elite universities, then reap the highest salaries – find that their reward is the unending pressure to work with 'crushing intensity' in order to maintain the income and status that have come to seem like prerequisites for the lives they want to lead.

It's not just that this situation feels impossible; in strictly logical terms, it really is impossible. It can't be the case that you *must* do more than you *can* do. That notion doesn't make any sense: if you truly don't have time for everything you want to do, or feel you ought to do, or that others are badgering you to do, then, well, you don't have time – no matter how grave the consequences of failing to do it all might prove to be. So, technically, it's irrational to feel troubled by an overwhelming to-do list. You'll do what you can, you won't do what you can't, and the tyrannical inner voice insisting that you must do everything is simply mistaken. We rarely stop to consider things so rationally, though, because that would mean confronting the painful truth of our limitations. We would be forced to acknowledge that there are hard choices to be made: which balls to let drop, which people to disappoint, which cherished ambitions to abandon, which roles to fail at. Maybe you *can't* keep your current job while also seeing enough of your children; maybe making sufficient time in the week for your creative calling means you'll *never* have an especially tidy home, or get quite as much exercise as you should, and so on. Instead, in an attempt to avoid these unpleasant truths, we deploy the strategy that dominates most conventional advice on how to deal with busyness: we tell ourselves we'll just have to find a way to do more – to try to address our busyness, you could say, by making ourselves busier still.

Sisyphus's Inbox

This is a modern reaction to a modern problem, but it is not brand new. In 1908, the English writer Arnold Bennett published a short and grouchy book of advice, the title of which demonstrated that this anxious effort to fit more in was already afflicting his Edwardian world: *How to Live on 24 Hours a Day*. 'Recently, in a daily organ, a battle raged round the question [of] whether a woman can exist nicely in the country on £85 a year,' Bennett wrote. 'I have [also] seen an essay, "How to live on eight shillings a week". But I have never seen an essay, "How to live on twenty-four hours a day".' The joke – to spell it out – is how absurd it is that anyone should need such advice, since nobody has ever had more than twenty-four hours a day in which to live. Yet people did need it: to Bennett and his target audience, suburban professionals commuting by tram and train to office jobs in England's increasingly prosperous cities, time was starting to feel like a container too small for all it was required to hold. He was writing, he explained, for his 'companions in distress – that innumerable band of souls who are haunted, more or less painfully, by the feeling that the years slip by, and slip by, and slip by, and that they have not yet been able to get their lives into proper working order'. His blunt diagnosis was that most people wasted several hours each day, especially in the evenings; they told themselves they were tired when they could just as easily pull up their socks and get on with all the life-enriching activities they claimed they never had time for. 'What I suggest,' wrote Bennett, 'is that at six o'clock you look facts in the face and admit that you are not tired (because you are not, you know).' As an alternative strategy, he suggests rising earlier instead; his book even contains instructions on how to brew your own tea, in case you're up before the servants.

How to Live on 24 Hours a Day is a wonderfully stimulating book, full of practical suggestions that make it well worth reading today. But the whole thing rests on one extremely dubious assumption (apart from the assumption that you have servants, I mean). Like virtually every time management expert who was to come after him, Bennett implies that if you follow his advice, you'll get enough of the genuinely important things done to feel at peace with time. Fit a bit more activity into each day's container, he suggests, and you'll reach the serene and commanding status of finally having 'enough time'. But that wasn't true in 1908, and it's even less true today. This was what I had begun to grasp, on that park bench in Brooklyn, and I still think it's the single best antidote to the feeling of time pressure, a splendidly liberating first step on the path of embracing your limits: the problem with trying to make time for everything that feels important – or just for *enough* of what feels important – is that you definitely never will.

The reason isn't that you haven't yet discovered the right time management tricks or applied sufficient effort, or that you need to start getting up earlier, or that you're generally useless. It's that the underlying assumption is unwarranted: there's no reason to believe you'll ever feel 'on top of things', or make time for everything that matters, simply by getting more done. For a start, what 'matters' is subjective, so you've no grounds for assuming that there will be time for everything that you, or your employer, or your culture happens to deem important. But the other exasperating issue is that if you succeed in fitting more in, you'll find the goalposts start to shift: more things will begin to seem important, meaningful, or obligatory. Acquire a reputation for doing your work at amazing speed, and you'll be given more of it. (Your boss isn't stupid: why would she give the extra work to someone slower?) Figure out how to spend enough time with your kids and at the office, so you don't feel guilty about either, and you'll suddenly feel some new social pressure: to spend more time exercising or to join the parent–teacher association – oh, and isn't it finally time you learned to meditate? Get around to launching the side business you've dreamed of for years, and if it succeeds, it won't be long before you're no longer satisfied with keeping it small. The same goes for chores: in her book *More Work for Mother*, the American historian Ruth Schwartz Cowan shows that when housewives first got access to 'labour-saving' devices like washing machines and vacuum cleaners, no time was saved at all, because society's standards of cleanliness simply rose to offset the benefits; now that you *could* return each of your husband's shirts to a spotless condition after a single wearing, it began to feel like you *should*, to show how much you loved him. 'Work expands so as to fill the time available for its completion,' the English humorist and historian C. Northcote Parkinson wrote in 1955, coining what became known as Parkinson's law. But it's not merely a joke, and it doesn't apply only to work. It applies to everything that needs doing. In fact, it's the definition of 'what needs doing' that expands to fill the time available.

This whole painful irony is especially striking in the case of email, that ingenious twentieth-century invention whereby any random person on the planet can pester you, at any time they like, and at almost no cost to themselves, by means of a digital window that sits inches from your nose, or in your pocket, throughout your working day, and often at weekends, too. The 'input' side of this arrangement – the number of emails that you could, in principle, receive – is essentially infinite. But the 'output' side – the number of messages you'll have time to read properly, reply to, or just make a considered decision to delete – is strictly finite. So getting better at processing your email is like getting faster and faster at climbing up an infinitely tall ladder: you'll feel more rushed, but no matter how quickly you go, you'll never reach the top. In ancient Greek myth, the gods punish King Sisyphus for his arrogance by sentencing him to push an enormous boulder up a hill, only to see it roll back down again, an action he is condemned to repeat for all eternity. In the contemporary version, Sisyphus would empty his inbox, lean back and take a deep breath, before hearing a familiar ping: 'You have new messages . . .'

It gets worse, though, because here the goalpost-shifting effect kicks in: every time you reply to an email, there's a good chance of provoking a reply to that email, which itself may require another reply, and so on and so on, until the heat death of the universe. At the same time, you'll become known as someone who responds promptly to email, so more people will consider it worth their while to message you to begin with. (By contrast, negligent emailers frequently find that forgetting to reply ends up saving them time: people find alternative solutions to the problems they were nagging you to solve, or the looming crisis they were emailing about never materialises.) So it's not simply that you never get through your email; it's that the process of 'getting through your email' actually *generates more email.* The general principle in operation is one you might call the 'efficiency trap'. Rendering yourself more efficient – either by implementing various productivity techniques or by driving yourself harder – won't generally result in the feeling of having 'enough time', because, all else being equal, the demands will increase to offset any benefits. Far from getting things done, you'll be creating new things to do.

For most of us, most of the time, it isn't feasible to avoid the efficiency trap altogether. After all, few of us are in a position *not* to attempt to get through our email, even if the consequence is that we receive more email. The same applies to life's other responsibilities, too: we're often obliged to find ways to cram more into the same amount of time, even if we end up feeling busier as a result. (Likewise, Schwartz Cowan's early-twentieth-century housewives presumably felt that they couldn't defy the social pressure towards ever tidier and cleaner homes.) So I don't mean to imply that once you grasp what's going on here, you'll magically never feel busy again.

But the choice you can make is to stop believing you'll ever solve the challenge of busyness by cramming more in, because that just makes matters worse. And once you stop investing in the idea that you might one day achieve peace of mind that way, it becomes easier to find peace of mind in the present, in the midst of overwhelming demands, because you're no longer making your peace of mind dependent on dealing with all the demands. Once you stop believing that it might somehow be possible to avoid hard choices about time, it gets easier to make better ones. You begin to grasp that when there's too much to do, and there always will be, the only route to psychological freedom is to let go of the limit-denying fantasy of getting it all done and instead to focus on doing a few things that count.

The Bottomless Bucket List

All this talk of inboxes and washing machines risks giving the impression that feeling overwhelmed is solely a matter of having too much to do at the office or around the house. But there's a deeper sense in which merely to be alive on the planet today is to be haunted by the feeling of having 'too much to do', whether or not you lead a busy life in any conventional sense. Think of it as 'existential overwhelm': the modern world provides an inexhaustible supply of things that seem worth doing, and so there arises an inevitable and unbridgeable gap between what you'd ideally like to do and what you actually can do. As the German sociologist Hartmut Rosa explains, pre-modern people weren't much troubled by such thoughts, partly because they believed in an afterlife: there was no particular pressure to 'get the most out of' their limited time, because as far as they were concerned, it wasn't limited, and in any case, earthly life was but a relatively insignificant prelude to the most important part. They also tended to see the world as unchanging through history or, in some cultures, as cycling repeatedly through the same predictable stages. It felt like a known quantity: you were content to play your role in the human drama – a role that countless thousands had played before you, and thousands more would play after your death – without any sense that you were missing out on the exciting new possibilities of your particular moment in history. (In an unchanging or cyclical view of history, there never are any exciting new possibilities.) But secular modernity changes all that. When people stop believing in an afterlife, everything depends on making the most of *this* life. And when people start believing in progress – in the idea that history is headed towards an ever more perfect future – they feel far more acutely the pain of their own little lifespan, which condemns them to missing out on almost all of that future. And so they try to quell their anxieties by cramming their lives with experience. In his translator's introduction to Rosa's book *Social Acceleration*, Jonathan Trejo-Mathys writes:

> The more we can accelerate our ability to go to different places, see new things, try new foods, embrace various forms of spirituality, learn new activities, share sensual pleasures with others whether it be in dancing or sex, experience different forms of art, and so on, the less incongruence there is between the possibilities of experience we can realize in our own lifetimes and the total array of possibilities available to human beings now and in the future – that is, the closer we come to having a truly 'fulfilled' life, in the literal sense of one that is as *filled full* of experiences as it can possibly be.

So the retiree ticking exotic destinations off a bucket list and the hedonist stuffing her weekends full of fun are arguably just as overwhelmed as the exhausted social worker or corporate lawyer. It's true that the things by which they're being overwhelmed are nominally more enjoyable; it's certainly nicer to have a long list of Greek islands left to visit than a long list of homeless families left to find housing for, or a huge stack of contracts left to proofread. But it remains the case that their fulfilment still seems to depend on their managing to do more than they *can* do. This helps explain why stuffing your life with pleasurable activities so often proves less satisfying than you'd expect. It's an attempt to devour the experiences the world has to offer, to feel like you've truly *lived* – but the world has an effectively infinite number of experiences to offer, so getting a handful of them under your belt brings you no closer to a sense of having feasted on life's possibilities. Instead, you find yourself pitched straight back into the efficiency trap. The more wonderful experiences you succeed in having, the more additional wonderful experiences you start to feel you could have, or ought to have, on top of all those you've already had, with the result that the feeling of existential overwhelm gets worse.

Perhaps it goes without saying that the internet makes this all much more agonising, because it promises to help you make better use of your time, while simultaneously exposing you to vastly more potential uses for your time – so that the very tool you're using to get the most out of life makes you feel as though you're missing out on even more of it. Facebook, for example, is an extremely efficient way to stay informed about events you might like to attend. But it's also a guaranteed way to hear about more events you'd like to attend than anyone possibly could attend. OkCupid is an efficient way of finding people to date, but also of being constantly reminded about all the other, potentially more alluring people you might be dating instead. Email is an unparalleled tool for responding rapidly to a large volume of messages – but then again, if it weren't for email, you wouldn't be receiving all those messages in the first place. The technologies we use to try to 'get on top of everything' always fail us, in the end, because they increase the size of the 'everything' of which we're trying to get on top.

Why You Should Stop Clearing the Decks

So far, I've been writing as if the efficiency trap were a simple matter of quantity: you have too much to do, so you try to fit more in, but the ironic result is that you end up with more to do. The worst aspect of the trap, though, is that it's also a matter of quality. The harder you struggle to fit everything in, the more of your time you'll find yourself spending on the least meaningful things. Adopt an ultra-ambitious time management system that promises to take care of your entire to-do list, and you probably won't even get round to the most important items on that list. Dedicate your retirement to seeing as much of the world as you possibly can, and you probably won't even get to see the most interesting parts.

The reason for this effect is straightforward: the more firmly you believe it ought to be possible to find time for everything, the less pressure you'll feel to ask whether any given activity is the best use for a portion of your time. Whenever you encounter some potential new item for your to-do list or your social calendar, you'll be strongly biased in favour of accepting it, because you'll assume you needn't sacrifice any other tasks or opportunities in order to make space for it. Yet because in reality your time is finite, doing anything requires sacrifice – the sacrifice of all the other things you could have been doing with that stretch of time. If you never stop to ask yourself if the sacrifice is worth it, your days will automatically begin to fill not just with more things, but with more trivial or tedious things, because they've never had to clear the hurdle of being judged more important than something else. Commonly, these will be things that other people want you to do, to make *their* lives easier, and which you didn't think to try to resist. The more efficient you get, the more you become 'a limitless reservoir for other people's expectations', in the words of the management expert Jim Benson.

In my days as a paid-up productivity geek, it was this aspect of the whole scenario that troubled me the most. Despite my thinking of myself as the kind of person who got things done, it grew painfully clear that the things I got done most diligently were the *unimportant* ones, while the important ones got postponed – either forever or until an imminent deadline forced me to complete them, to a mediocre standard and in a stressful rush. The email from my newspaper's IT department about the importance of regularly changing my password would provoke me to speedy action, though I could have ignored it entirely. (The clue was in the subject line, where the words 'PLEASE READ' are generally a sign you needn't bother reading what follows.) Meanwhile, the long message from an old friend now living in New Delhi and research for the major article I'd been planning for months would get ignored, because I told myself that such tasks needed my full focus, which meant waiting until I had a good chunk of free time and fewer small-but-urgent tasks tugging at my attention. And so, instead, like the dutiful and efficient worker I was, I'd put my energy into clearing the decks, cranking through the smaller stuff to get it out of the way – only to discover that doing so took the whole day, that the decks filled up again overnight anyway, and that the moment for responding to the New Delhi email or for researching the milestone article never arrived. One can waste years this way, systematically postponing precisely the things one cares about the most.

What's needed instead in such situations, I gradually came to understand, is a kind of anti-skill: not the counter-productive strategy of trying to make yourself more efficient, but rather a willingness to resist such urges – to learn to stay with the anxiety of feeling overwhelmed, of not being on top of everything, without automatically responding by trying to fit more in. To approach your days in this fashion means, instead of clearing the decks, *declining* to clear the decks, focusing instead on what's truly of greatest consequence while tolerating the discomfort of knowing that, as you do so, the decks will be filling up further, with emails and errands and other to-dos, many of which you may never get round to at all. You'll sometimes still decide to drive yourself hard in an effort to squeeze more in, when circumstances absolutely require it. But that won't be your default mode, because you'll no longer be operating under the illusion of one day making time for everything.

The same goes for existential overwhelm: what's required is the will to resist the urge to consume more and more experiences, since that strategy can only lead to the feeling of having even more experiences left to consume. Once you truly understand that you're guaranteed to miss out on almost every experience the world has to offer, the fact that there are so many you still haven't experienced stops feeling like a problem. Instead, you get to focus on fully enjoying the tiny slice of experiences you actually do have time for – and the freer you are to choose, in each moment, what counts the most.

The Pitfalls of Convenience

There's one further, especially insidious way in which the quest for increased efficiency warps our relationship with time these days: the seductive lure of *convenience*. Entire industries now thrive on the promise of helping us cope with having an overwhelming amount to do by eliminating or accelerating tedious and time-consuming chores. But the result – in an irony that shouldn't be too surprising by now – is that life gets subtly worse. As with other manifestations of the efficiency trap, freeing up time in this fashion backfires in terms of quantity, because the freed-up time just fills with more things you feel you have to do, and also in terms of quality, because in attempting to eliminate only the tedious experiences, we accidentally end up eliminating things we didn't realise we valued until they were gone.

It works like this: in start-up jargon, the way to make a fortune in Silicon Valley is to identify a 'pain point' – one of the small annoyances resulting from (more jargon) the 'friction' of daily life – and then to offer a way to circumvent it. Thus Uber eliminates the 'pain' of having to track down a number for your local taxi company and call it, or trying to hail a cab in the street; digital wallet apps like Apple Pay remove the 'pain' of having to reach into your bag for your physical wallet or cash. The US food delivery service Seamless has even run advertisements – tongue-in-cheek ones, but still – boasting that it lets you avoid the agony of talking to a flesh-and-blood restaurant worker; instead, you need only commune with a screen. It's true that everything runs more smoothly this way. But smoothness, it turns out, is a dubious virtue, since it's often the unsmoothed textures of life that make it liveable, helping nurture the relationships that are crucial for mental and physical health, and for the resilience of our communities. Your loyalty to your local taxi firm is one of those delicate social threads that, multiplied thousands of times, bind a neighbourhood together; your interactions with the woman who runs the nearby Chinese takeaway might feel insignificant, but they help make yours the kind of area where people still talk to one another, where tech-induced loneliness doesn't yet reign supreme. (Take it from a work-from-home writer: a couple of brief interactions with another human can make all the difference in a day.) As for Apple Pay, I like a little friction when I buy something, since it marginally increases the chance that I'll resist a pointless purchase.

Convenience, in other words, makes things easy, but without regard to whether easiness is truly what's most valuable in any given context. Take those services – on which I've relied too much in recent years – that let you design and then remotely mail a birthday card, so you never see or touch the physical item yourself. Better than nothing, perhaps. But sender and recipient both know that it's a poor substitute for purchasing a card in a shop, writing on it by hand, and then walking to a postbox to post it, because contrary to the cliché, it isn't really the thought that counts, but the effort – which is to say, the inconvenience. When you render the process more convenient, you drain it of its meaning. The venture capitalist and Reddit co-founder Alexis Ohanian has observed that we often 'don't even realize something is broken until someone else shows us a better way'. But the other reason we might not realise some everyday process is broken is that it isn't broken to begin with – and that the inconvenience involved, which might look like brokenness from the outside, in fact embodies something essentially human.

Frequently, the effect of convenience isn't just that a given activity starts to feel less valuable, but that we stop engaging in certain valuable activities altogether, in favour of more convenient ones. Because you *can* stay home, order food online, and watch sitcoms on Netflix, you find yourself doing so – though you might be perfectly well aware that you'd have had a better time had you kept your appointment to meet friends in town or tried to make an interesting new recipe. 'I prefer to brew my coffee,' the Colombia University law professor Tim Wu writes in an essay on the pitfalls of convenience culture, 'but Starbucks instant is so convenient I hardly ever do what I "prefer."' Meanwhile, those aspects of life that resist being made to run more smoothly start to seem repellent. 'When you can skip the line and buy concert tickets on your "phone",' Wu points out, 'waiting in line to vote in an election is irritating.' As convenience colonises everyday life, activities gradually sort themselves into two types: the kind that are now far more convenient, but that feel empty or out of sync with our true preferences; and the kind that now seem intensely annoying, because of how inconvenient they remain.

Resisting all this as an individual, or as a family, takes fortitude, because the smoother life gets, the more perverse you'll seem if you insist on maintaining the rough edges by choosing the inconvenient way of doing things. Get rid of your smartphone, stop using Google, or choose snail mail over WhatsApp, and people are increasingly likely to question your sanity. Still, it can be done. The Bible scholar and agriculturalist Sylvia Keesmaat abandoned a full-time university position in Toronto because she was following a hunch that her overwhelmed life – and the efficiencies and conveniences it seemed to necessitate – were somehow undermining its meaning. She moved with her husband and children to a farm in the vast swathe of the Canadian interior known as the Land Between, where each winter day begins by lighting the fire that will warm the farmhouse and provide heat for cooking:

> Every morning I carefully scrape out the ash of yesterday . . . As I lay the kindling and listen for the crackling of wood devouring flame, I wait. The house is cool, and all I have to do now for the next few minutes is be attentive and patient. The fire needs time to build, needs to be fed and nurtured into the strength of heat for cooking. If I walk away and leave it, it will die. If I forget to pay attention, it will die. Of course, being fire, if I build it too big and forget to pay attention, *I* could die. Why take the chance?
>
> Someone once asked me how long it takes before I have my first, hot cup of tea in the morning. Well, let's see: in the winter I light the fire, sweep the floor, and wake the kids for chores . . . I run water for the cows, get them some hay, give the chickens some grain and their water, feed the ducks. Sometimes I help the kids with

> the horses and barn cats and then come back in. Then I put the kettle on. Maybe I get something to drink within an hour of waking. If things go well. An hour?

We needn't dwell here on whether Keesmaat's new, self-consciously inconvenient way of life is intrinsically superior to the kind with central heating, takeaways and twice-daily commutes. (Although I think perhaps it might be: her days do seem busy in exactly that agreeable, non-overwhelmed, Richard Scarry sense of the word.) And obviously not everyone has the option of pursuing precisely her sort of path. But the real point is that her decision to make such a radical change arose from the recognition that she'd never manage to build a more meaningful life – which for her meant cultivating a more mindful relationship with her family's physical surroundings – by saving time and thereby squeezing more into her existing one. To make time for what mattered, she needed to give things up.

Convenience culture seduces us into imagining that we might find room for everything important by eliminating only life's tedious tasks. But it's a lie. You have to choose a few things, sacrifice everything else, and deal with the inevitable sense of loss that results. Keesmaat chose building fires and growing food with her children. 'How else are we to get to know this place where we have been set, apart from tending to it?' she writes. 'Outside of planting the food we eat, how are we to learn the living character of soil, the various needs of peppers, lettuce, and kale?' You might make a very different choice, of course. But the undodgeable reality of a finite human life is that you *are* going to have to choose.

3.

Facing Finitude

You can't delve far into the question of what it means to be a finite human being, with finite time on the planet, before encountering the philosopher who was more obsessed with the subject than any other thinker: Martin Heidegger. This is unfortunate for two reasons, the most glaring one being that for more than a decade, starting in 1933, he was a card-carrying member of the Nazi Party. (The question of what this means for his philosophy is a fraught and fascinating one, but it would get us off track here. So you're going to have to decide for yourself whether this exceptionally poor life choice invalidates his thoughts about how we make life choices in general.) The second reason is that he's almost impossible to read. His work abounds with broken-backed phrases such as 'Being-towards-death' and 'de-severance' and – you might like to be sitting down for this next one – 'anxiety "in the face of" that potentiality-for-Being which is one's ownmost'. This is why nobody's interpretation of Heidegger's work, very much including mine, ought to be taken as definitive. Yet on this second charge, of incomprehensibility, he does have a kind of defence. Everyday language reflects our everyday ways of seeing. But Heidegger wants to slide his fingernails under the most basic elements of existence – the things we barely notice because they're so familiar – so as to prise them away for our inspection. That means making things unfamiliar, using unfamiliar terms. So you stumble and trip over his writing, but sometimes, as a consequence, you bang your head against reality.

Thrown into Time

The most fundamental thing we fail to appreciate about the world, Heidegger asserts in his magnum opus, *Being and Time*, is how bafflingly astonishing it is that it's there at all – the fact that there is anything rather than nothing. Most philosophers and scientists spend their careers pondering the *way* things are: what sorts of things exist, where they come from, how they relate to each other, and so on. But we've forgotten to be amazed *that* things are in the first place – that 'a world is worlding all around us', as Heidegger puts it. This fact – the fact that *there is being*, to begin with – is 'the brute reality on which all of us ought to be constantly stubbing our toes', in the splendid phrase of the writer Sarah Bakewell. But instead, it almost always passes us by.

Having focused our attention on this ground-level issue

of 'being' itself, Heidegger next turns to humans specifically, and to our own particular kind of being. What does it mean for a human being to *be*? (I realise this is starting to sound like a bad comedy sketch about philosophers lost in wild abstractions. I'm afraid that's going to get worse for another couple of paragraphs before it gets better.) His answer is that our being is totally, utterly bound up with our finite time. So bound up, in fact, that the two are synonymous: to be, for a human, is above all to exist temporally, in the stretch between birth and death, certain that the end will come, yet unable to know when. We tend to speak about our *having* a limited amount of time. But it might make more sense, from Heidegger's strange perspective, to say that we *are* a limited amount of time. That's how completely our limited time defines us.

Ever since Heidegger made this claim, philosophers have been disagreeing about what exactly it might mean to say that we are time – some have even argued it doesn't mean anything – so we shouldn't get stuck on trying to clarify it with precision. It's sufficient to take from it the insight that every moment of a human existence is completely shot through with the fact of what Heidegger calls our 'finitude'. Our limited time isn't just one among various things we have to cope with; rather, it's the thing that defines us, as humans, before we start coping with anything at all. Before I can ask a single question about what I should do with my time, I find myself already thrown *into* time, into this particular moment, with my particular life story, which has made me who I am and which I can never get out from under. Looking ahead to the future, I find myself equally constrained by my finitude: I'm being borne forward on the river of time, with no possibility of stepping out of the flow, onward towards my inevitable death – which, to make matters even more ticklish, could arrive at any moment.

In this situation, any decision I make, to do anything at all with my time, is already radically limited. For one thing, it's limited in a retrospective sense, because I'm already who I am and where I am, which determines what possibilities are open to me. But it's also radically limited in a forward-looking sense, too, not least because a decision to do any given thing will automatically mean sacrificing an infinite number of potential alternative paths. As I make hundreds of small choices throughout the day, I'm building a life – but at one and the same time, I'm closing off the possibility of countless others, forever. (The original Latin word for 'decide', *decidere*, means 'to cut off', as in slicing away alternatives; it's a close cousin of words like 'homicide' and 'suicide'.) Any finite life – even the best one you could possibly imagine – is therefore a matter of ceaselessly waving goodbye to possibility.

The only real question about all this finitude is whether we're willing to confront it or not. And this, for Heidegger, is the central challenge of human existence: since finitude defines our lives, he argues that living a truly authentic life – becoming fully human – means facing up to that fact. We must live out our lives, to whatever extent we can, in clear-eyed acknowledgement of our limitations, in the undeluded mode of existence that Heidegger calls Being-towards-death', aware that *this is it*, that life is not a dress rehearsal, that every choice requires myriad sacrifices, and that time is always already running out – indeed, that it may run out today, tomorrow, or next month. And so it's not merely a matter of spending each day 'as if' it were your last, as the cliché has it. The point is that it always actually might be. I can't entirely depend upon a single moment of the future.

Obviously, from any ordinary perspective, this all sounds intolerably morbid and stressful. But then, to the extent that you manage to achieve this outlook on life, you're *not* seeing it from an ordinary perspective – and 'morbid and stressful', at least according to Heidegger, are exactly what it is not. On the contrary, it's the only way for a finite human being to live fully, to relate to other people as full-fledged humans, and to experience the world as it truly is. What's really morbid, from this perspective, is what most of us do, most of the time, instead of confronting our finitude, which is to indulge in avoidance and denial, or what Heidegger calls 'falling'. Rather than taking ownership of our lives, we seek out distractions, or lose ourselves in busyness and the daily grind, so as to try to forget our real predicament. Or we try to avoid the intimidating responsibility of having to decide what to do with our finite time by telling ourselves that we don't get to choose at all – that we must get married, or remain in a soul-destroying job, or anything else, simply because it's the done thing. Or, as we saw in the previous chapter, we embark on the futile attempt to 'get everything done', which is really another way of trying to evade the responsibility of deciding what to do with your finite time – because if you actually *could* get everything done, you'd never have to choose among mutually exclusive possibilities. Life is usually more comfortable when you spend it avoiding the truth in this fashion. But it's a stultifying, deadly sort of comfort. It's only by facing our finitude that we can step into a truly authentic relationship with life.

Getting Real

In his 2019 book, *This Life*, the Swedish philosopher Martin Hägglund makes this all a bit clearer and less mystical by juxtaposing the idea of facing our finitude with the religious belief in an eternal life. If you really thought life would never end, he argues, then nothing could ever genuinely matter, because you'd never be faced with having to decide whether or not to use a portion of your precious life on something. 'If I believed that my life would last forever,' Hägglund writes, 'I could never take my life to be at stake, and I would never be seized by the need to do anything with my time.' Eternity would be deathly dull, because whenever you found yourself wondering whether or not to do any given thing, on any given day, the answer would always be: Who cares? After all, there's always tomorrow, and the next day, and the one after that . . . Hägglund quotes a headline from the magazine *U.S. Catholic* that has the air of being written by a devout religious believer on whom an awful possibility has suddenly dawned: 'Heaven: Will It Be Boring?'

By way of contrast, Hägglund describes the annual summer holiday he spends with his extended family in a house on Sweden's wind-battered Baltic coast. It's intrinsic to the value of this experience, he notes, that he won't be around to experience it forever, that his relatives won't either, that his relationships with his relatives are therefore temporary, too – and that even the coastline, in its current form, is a transient phenomenon, as dry land continues to emerge from the 12,000-year retreat of the region's glaciers. If Hägglund were guaranteed an infinity of these summer holidays, there'd be nothing much to value about any one of them; it's only the guarantee that he definitely won't have an infinity of them that makes them worth valuing. Indeed, it's also only from this position of valuing what is finite because it's finite, Hägglund argues, that one can truly care about the impact of a collective peril such as climate change, which is wreaking changes to his native country's landscape. If our earthly existence were merely the prelude to an eternity in heaven, threats to that existence couldn't matter in any ultimate sense.

Of course, if you're not religious, and maybe even if you are, you might not literally believe in eternal life. But anyone who spends their days failing to confront the truth of their finitude – convincing themselves, on a subconscious level, that they have all the time in the world, or alternatively that they'll be able to cram an infinite amount into the time they do have – is essentially in the same boat. They're living in denial of the fact that their time is limited; so when it comes to deciding how to use any given portion of that time, nothing can genuinely be at stake for them. It is by consciously confronting the certainty of death, and what follows from the certainty of death, that we finally become truly present for our lives.

This is the kernel of wisdom in the cliché of the celebrity who claims that a brush with cancer was 'the best thing that ever happened' to them: it pitches them into a more authentic mode of being, in which everything suddenly feels more vividly meaningful. Such accounts sometimes give the impression that people reliably become happier as a result of facing the truth about death, which isn't the case; 'happier' is clearly the wrong word for the new depth that is added to life when you grasp, deep in your bones, the fact that you're going to die and that your time is therefore severely limited. But things certainly do get *realer*. As she recalls in her memoir *The Iceberg*, the sculptor Marion Coutts was taking her two-year-old son to his first day with a new childminder when her husband, the art critic Tom Lubbock, came to find her to tell her about the malignant brain tumour from which he was to die within three years:

> Something has happened. A piece of news. We have had a diagnosis that has the status of an event. The news makes a rupture with what went before: clean, complete and total, save in one respect. It seems that after the event, the decision we make is to remain. Our [family] unit stands . . .
>
> We learn something. We are mortal. You might say you know this but you don't. The news falls neatly between one moment and another. You would not think there was a gap for such a thing . . . It is as if a new physical law has been described for us bespoke: absolute as all the others are, yet terrifyingly casual. It is a law of perception. It says, *You will lose everything that catches your eye.*

In case this needs saying, it isn't that a diagnosis of terminal illness, or a bereavement, or any other encounter with death is somehow good, or desirable, or 'worth it'. But such experiences, however wholly unwelcome, often appear to leave those who undergo them in a new and more honest relationship with time. The question is whether we might attain at least a little of that same outlook in the absence of the experience of agonising loss. Writers have struggled to convey the particular quality that this mode of being infuses into life, because while 'happier' is wrong, 'sadder' doesn't convey it, either. You might call it 'bright sadness' (as does the priest and author Richard Rohr), 'stubborn gladness' (the poet Jack Gilbert), or 'sober joy' (the Heidegger scholar Bruce Ballard). Or you could just call it finally encountering real life, and the brute fact of our finite weeks.

Everything Is Borrowed Time

This is the point at which I should come clean and admit that, unfortunately, I don't live my own daily life in a permanent state of unflinching acceptance of my mortality. Perhaps nobody does. What I can confirm, though, is that if you can adopt the outlook we're exploring here even just a little – if you can hold your attention, however briefly or occasionally, on the sheer astonishingness of *being*, and on what a small amount of that being you get – you may experience a palpable shift in how it feels to be here, right now, alive in the flow of time. (Or *as* the flow of time, a Heideggerian might say.) From an everyday standpoint, the fact that life is finite feels like a terrible insult, 'a sort of personal affront, a taking-away of one's time', in the words of one scholar. There you were, planning to live on forever – as the old Woody Allen line has it, not in the hearts of your countrymen, but in your apartment – but now here comes mortality, to steal away the life that was rightfully yours.

Yet, on reflection, there's something very entitled about this attitude. Why assume that an infinite supply of time is the default, and mortality the outrageous violation? Or to put it another way, why treat four thousand weeks as a very small number, because it's so tiny compared with infinity, rather than treating it as a huge number, because it's so many more weeks than if you had never been born? Surely only somebody who'd failed to notice how remarkable it is that anything *is*, in the first place, would take their own being as such a given – as if it were something they had every right to have conferred upon them, and never to have taken away. So maybe it's not that you've been cheated out of an unlimited supply of time; maybe it's almost incomprehensibly miraculous to have been granted any time at all.

The Canadian writer David Cain understood all this with a jolt in the summer of 2018 when he attended an event in the Greektown district of Toronto. The evening itself passed off unremarkably: 'I was early,' he recalled, 'so I spent some time in a nearby park, then checked out the shops and restaurants on Danforth Avenue. I stopped in front of a church to tie my shoe. I remember being nervous about meeting a bunch of new people.' Then, two weeks later, on the same stretch of street, a deranged man shot fourteen people, killing two of them, then killing himself. Rationally speaking, Cain concedes, this wasn't a narrow escape on his part; thousands of people walk down Danforth Avenue every day, and it wasn't as if he'd missed the shooting by only a few minutes. Even so, the sense that it could have been him caught in that gunfire was sufficiently powerful to bring into focus what it meant that it hadn't been him. 'When I watched videos of eye-witness accounts, including some in front of the church where I tied my shoes and the corner where I nervously loitered,' he wrote later, 'it gave me a vital bit of perspective: I happen to be alive, and there's no cosmic law entitling me to that status. Being alive is just happenstance, and not one more day of it is guaranteed.'

This kind of perspective shift, I've found, has an especially striking effect on the experience of everyday annoyances – on my response to traffic jams and airport security lines, babies who won't sleep past 5 a.m., and dishwashers that I apparently must empty again tonight, even though (I think you'll find!) I did so yesterday. I'm embarrassed to admit what an outsize negative effect such minor frustrations have had on my happiness over the years. Fairly often, they still do; but the effect was worst at the height of my productivity geekhood, because when you're trying to Master Your Time, few things are more infuriating than a task or delay that's foisted upon you against your will, with no regard for the schedule you've painstakingly drawn up in your overpriced notebook. But when you turn your attention instead to the fact that you're in a position to *have* an irritating experience in the first place, matters are liable to look very different indeed. All at once, it can seem amazing to be there at all, having any experience, in a way that's overwhelmingly more important than the fact that the experience happens to be an annoying one. Geoff Lye, an environmental consultant, once told me that after the sudden and premature death of his friend and colleague David Watson, he would find himself stuck in traffic, not clenching his fists in agitation, as per usual, but wondering: 'What would David have given to be caught in this traffic jam?' It was the same for queues in supermarkets and customer service lines that kept him on hold too long. Lye's focus was no longer exclusively on *what* he was doing in such moments or what he'd rather be doing instead; now, he noticed also *that* he was doing it, with an upwelling of gratitude that took him by surprise.

And now consider what all this means for the crucial and basic question of choosing what to do with your limited time. As we've seen, it's a fact of life that, as a finite human, you're always making hard choices – so that, for example, in spending this afternoon on one thing that mattered to me (writing), I necessarily had to forgo many other things that mattered too (like playing with my son). It's natural to see this situation as highly regrettable, and to yearn for some alternative version of existence in which we wouldn't have to choose between valued activities in this way. But if it's amazing to have been granted any being at all – if 'your whole life is borrowed time', as Cain realised, watching news reports of the Danforth Avenue shootings – then wouldn't it make more sense to speak not of having to make such choices, but of getting to make them? From this viewpoint, the situation starts to seem much less regrettable: each moment of decision becomes an opportunity to select from an enticing menu of possibilities, when you might easily never have been presented with the menu to begin with. And it stops making sense to pity yourself for having been cheated of all the other options.

In this situation, making a choice – picking one item

from the menu – far from representing some kind of defeat, becomes an affirmation. It's a positive commitment to spend a given portion of time doing *this* instead of *that* – actually, instead of an infinite number of other 'thats' – because *this*, you've decided, is what counts the most right now. In other words, it's precisely the fact that I could have chosen a different and perhaps equally valuable way to spend this afternoon that bestows meaning on the choice I did make. And the same applies, of course, to an entire lifetime. For instance, it's precisely the fact that getting married forecloses the possibility of meeting someone else – someone who might genuinely have been a better marriage partner; who could ever say? – that makes marriage meaningful. The exhilaration that sometimes arises when you grasp this truth about finitude has been called the 'joy of missing out', by way of a deliberate contrast with the idea of the 'fear of missing out'. It is the thrilling recognition that you wouldn't even really want to be able to do everything, since if you didn't have to decide what to miss out on, your choices couldn't truly mean anything. In this state of mind, you can embrace the fact that you're forgoing certain pleasures, or neglecting certain obligations, because whatever you've decided to do instead – earn money to support your family, write your novel, bath the toddler, pause on a hiking trail to watch a pale winter sun sink below the horizon at dusk – is how you've chosen to spend a portion of time that you never had any right to expect.

4.

Becoming a Better Procrastinator

Perhaps we're in danger of getting a little too metaphysical about all this, though. Many of the philosophers who've pondered the subject of human finitude have been reluctant to translate their observations into practical advice, because that smacks of self-help. (And heaven forbid that anyone might want to help themselves!) Yet their insights do have concrete ramifications for daily life. Apart from anything else, they make it clear that the core challenge of managing our limited time isn't about how to get everything done – that's never going to happen – but how to decide most wisely what *not* to do, and how to feel at peace about not doing it. As the American author and teacher Gregg Krech puts it, we need to learn to get better at procrastinating. Procrastination of some kind is inevitable:

indeed, at any given moment, you'll be procrastinating on almost everything, and by the end of your life, you'll have got round to doing virtually none of the things you theoretically could have done. So the point isn't to eradicate procrastination, but to choose more wisely what you're going to procrastinate on, in order to focus on what matters most. The real measure of any time management technique is *whether or not it helps you neglect the right things*.

A large proportion of them don't. They make matters worse. Most productivity experts act merely as enablers of our time troubles, by offering ways to keep on believing it might be possible to get everything done. Perhaps you're familiar with the extraordinarily irritating parable of the rocks in the jar, which was first inflicted upon the world in Stephen Covey's 1994 book, *First Things First*, and which has been repeated ad nauseam in productivity circles ever since. In the version with which I'm most familiar, a teacher arrives in class one day carrying several sizeable rocks, some pebbles, a bag of sand and a large glass jar. He issues a challenge to his pupils: Can they fit all the rocks, pebbles and sand into the jar? The pupils, who are apparently rather slow-witted, try putting the pebbles or the sand in first, only to find that the rocks won't fit. Eventually – and no doubt with a condescending smile – the teacher demonstrates the solution: he puts the rocks in first, then the pebbles, then the sand, so that the smaller items nestle comfortably in the spaces between the larger ones. The moral is that if you make time for the most important things first, you'll get them all done and have plenty of room for less important things besides. But if you don't approach your to-do list in this order, you'll never fit the bigger things in at all.

Here the story ends – but it's a lie. The smug teacher is being dishonest. He has rigged his demonstration by bringing only a few big rocks into the classroom, knowing they'll all fit into the jar. The real problem of time management today, though, isn't that we're bad at prioritising the big rocks. It's that there are too many rocks – and most of them are never making it anywhere near that jar. The critical question isn't how to differentiate between activities that matter and those that don't, but what to do when far too many things feel at least somewhat important, and therefore arguably qualify as big rocks. Fortunately, a handful of wiser minds have addressed exactly this dilemma, and their counsel coalesces around three main principles.

The Art of Creative Neglect

Principle number one is to *pay yourself first* when it comes to time. I'm borrowing this phrasing from the graphic novelist and creativity coach Jessica Abel, who borrowed it in turn from the world of personal finance, where it's long been an article of faith because it works. If you take a portion of your pay cheque the day you receive it and squirrel it away into savings or investments, or use it for paying off debts, you'll probably never feel the absence of that cash; you'll go about your business – buying your groceries, paying your bills – precisely as if you'd never had that portion of money to begin with. (There are limits, of course: this plan won't work if you literally earn only enough to survive.) But if, like most people, you 'pay yourself last' instead – buying what you need and hoping there'll be some money remaining at the end to put into savings – you'll usually find that there isn't any. And this won't necessarily be because you frittered it away self-indulgently, on lattes, or pedicures, or new electronic gadgets, or heroin. Every expenditure might have felt eminently sensible and necessary in the moment that you made it. The trouble is that we're terrible at long-range planning: if something feels like a priority now, it's virtually impossible to coolly assess whether it will still feel that way in a week or a month. And so we naturally err on the side of spending – then feel bad later when there's nothing left over to save.

The same logic, Abel points out, applies to time. If you try to find time for your most valued activities by first dealing with all the other important demands on your time, in the hope that there'll be some left over at the end, you'll be disappointed. So if a certain activity really matters to you – a creative project, say, though it could just as easily be nurturing a relationship, or activism in the service of some cause – the only way to be sure it will happen is to do some of it today, no matter how little, and no matter how many other genuinely big rocks may be begging for your attention. After years of trying and failing to make time for her illustration work, by taming her to-do list and shuffling her schedule, Abel saw that her only viable option was to claim time instead – to just start drawing, for an hour or two, every day, and to accept the consequences, even if those included neglecting other activities she sincerely valued. 'If you don't save a bit of your time for you, now, out of every week,' as she puts it, 'there is no moment in the future when you'll magically be done with everything and have loads of free time.' This is the same insight embodied in two venerable pieces of time management advice: to work on your most important project for the first hour of each day, and to protect your time by scheduling 'meetings' with yourself, marking them in your calendar so that other commitments can't intrude. Thinking in terms of 'paying yourself first' transforms these one-off tips into a philosophy of life, at the core of which lies this simple insight: if you plan to spend some of your four thousand weeks doing what matters most to you, then at some point you're just going to have to start doing it.

The second principle is to *limit your work in progress*. Perhaps the most appealing way to resist the truth about your finite time is to initiate a large number of projects at once; that way, you get to feel as though you're keeping plenty of irons in the fire and making progress on all fronts. Instead, what usually ends up happening is that you make progress on no fronts – because each time a project starts to feel difficult, or frightening, or boring, you can bounce off to a different one instead. You get to preserve your sense of being in control of things, but at the cost of never finishing anything important.

The alternative approach is to fix a hard upper limit on the number of things that you allow yourself to work on at any given time. In their book *Personal Kanban*, which explores this strategy in detail, the management experts Jim Benson and Tonianne DeMaria Barry suggest no more than three items. Once you've selected those tasks, all other incoming demands on your time must wait until one of the three items has been completed, thereby freeing up a slot. (It's also permissible to free up a slot by abandoning a project altogether if it isn't working out. The point isn't to force yourself to finish absolutely everything you start, but rather to banish the bad habit of keeping an ever-proliferating number of half-finished projects on the back burner.)

Making this rather modest change to my working practices produced a startlingly large effect. It was no longer possible for me to ignore the fact that my capacity for work was strictly finite – because each time I selected a new task from my to-do list, as one of my three work-in-progress items, I was obliged to contemplate all those I'd inevitably be neglecting in order to focus on it. And yet precisely because I was being forced to confront reality in this way – to see that I was *always* neglecting most tasks, in order to work on anything at all, and that working on everything at once simply wasn't an option – the result was a powerful sense of undistracted calm, and a lot more productivity than in my days as a productivity obsessive. Another happy consequence was that I found myself effortlessly breaking down my projects into manageable chunks, a strategy I'd long agreed with in theory but never properly implemented. Now it became the intuitive thing to do: it was clear that if I nominated 'write book' or 'move house' as one of my three tasks in progress, it would clog up the system for months, so I was naturally motivated to figure out the next achievable step in each case instead. Rather than trying to do everything, I found it easier to accept the truth that I'd be doing only a few things on any given day. The difference, this time, was that I actually did them.

The third principle is to *resist the allure of middling priorities*. There is a story attributed to Warren Buffett – although probably only in the apocryphal way in which wise insights get attributed to Albert Einstein or the Buddha, regardless of their real source – in which the famously shrewd investor is asked by his personal pilot about how to set priorities. I'd be tempted to respond, 'Just focus on flying the plane!' But apparently this didn't take place mid-flight, because Buffett's advice is different: he tells the man to make a list of the top twenty-five things he wants out of life and then to arrange them in order, from the most important to the least. The top five, Buffett says, should be those around which he organises his time. But contrary to what the pilot might have been expecting to hear, the remaining twenty, Buffett allegedly explains, aren't the second-tier priorities to which he should turn when he gets the chance. Far from it. In fact, they're the ones he should actively avoid at all costs – because they're the ambitions insufficiently important to him to form the core of his life yet seductive enough to distract him from the ones that matter most.

You needn't embrace the specific practice of listing out your goals (I don't, personally) to appreciate the underlying point, which is that in a world of too many big rocks, it's the moderately appealing ones – the fairly interesting job opportunity, the semi-enjoyable friendship – on which a finite life can come to grief. It's a self-help cliché that most of us need to get better at learning to say no. But as the writer

Elizabeth Gilbert points out, it's all too easy to assume that this merely entails finding the courage to decline various tedious things you never wanted to do in the first place. In fact, she explains, 'it's much harder than that. You need to learn how to start saying no to things you *do* want to do, with the recognition that you have only one life.'

Perfection and Paralysis

If skilful time management is best understood as a matter of learning to procrastinate well, by facing the truth about your finitude and making your choices accordingly, then the *other* kind of procrastination – the bad kind, which prevents us from making progress on the work that matters to us – is usually the result of trying to avoid that truth. The good procrastinator accepts the fact that she can't get everything done, then decides as wisely as possible what tasks to focus on and what to neglect. By contrast, the bad procrastinator finds himself paralysed precisely because he can't bear the thought of confronting his limitations. For him, procrastination is a strategy of emotional avoidance – a way of trying not to feel the psychological distress that comes with acknowledging that he's a finite human being.

The limitations we're trying to avoid when we engage in this self-defeating sort of procrastination frequently don't have anything to do with how *much* we'll be able to get done in the time available; usually, it's a matter of worrying that we won't have the talent to produce work of sufficient quality, or that others won't respond to it as we'd like them to, or that in some other way things won't turn out as we want. The philosopher Costica Bradatan illustrates the point by means of a fable about an architect from Shiraz in Persia who designed the world's most beautiful mosque: a breathtaking structure, dazzlingly original yet classically well proportioned, awe-inspiring in its grandeur yet wholly unpretentious. All those who saw the architectural plans wanted to buy them, or steal them; famous builders begged him to let them take on the job. But the architect locked himself in his study and stared at the plans for three days and nights – then burned them all. He might have been a genius, but he was also a perfectionist: the mosque of his imagination was perfect, and it agonised him to contemplate the compromises that would be involved in making it real. Even the greatest of builders would inevitably fail to reproduce his plans absolutely faithfully; nor would he be able to protect his creation from the ravages of time – from the physical decay or marauding armies that would eventually reduce it to dust. Stepping into the world of finitude, by actually building the mosque, would mean confronting all that he couldn't do. Better to cherish an ideal fantasy than to resign himself to reality, with all its limitations and unpredictability.

Bradatan argues that when we find ourselves procrastinating on something important to us, we're usually in some version of this same mindset. We fail to see, or refuse to accept, that any attempt to bring our ideas into concrete reality must inevitably fall short of our dreams, no matter how brilliantly we succeed in carrying things off – because reality, unlike fantasy, is a realm in which we don't have limitless control, and can't possibly hope to meet our perfectionist standards. Something – our limited talents, our limited time, our limited control over events and over the actions of other people – will always render our creation less than perfect. Dispiriting as this might sound at first, it contains a liberating message: if you're procrastinating on something because you're worried you won't do a good enough job, you can relax – because judged by the flawless standards of your imagination, you definitely *won't* do a good enough job. So you might as well make a start.

And this sort of finitude-avoiding procrastination certainly isn't confined to the world of work. It's a major issue in relationships, too, where a similar refusal to face the truth about finitude can keep people mired in a miserably tentative mode of existence for years on end. By way of a cautionary tale, consider the case of the worst boyfriend ever, Franz Kafka, whose most important romantic liaison began one summer evening in Prague in 1912, when he was twenty-nine. Dining that night at the home of his friend Max Brod, Kafka met his host's cousin, Felice Bauer, who was visiting from Berlin. She was an independent-minded twenty-four-year-old, already enjoying professional success at a manufacturing company in Germany, and her down-to-earth vigour appealed to the neurotic, self-conscious Kafka. We know little of the strength of feeling in the other direction, since only Kafka's account survives, but he was smitten, and soon a relationship began.

Or it began, at least, in correspondence form: over the next five years, the couple exchanged hundreds of letters yet met only a handful of times, each meeting apparently a source of agony for Kafka. Seven months after their first encounter, he finally agreed to meet a second time, but sent a telegram on the morning in question to say he wasn't coming; then he showed up anyway but acted morose. When the couple eventually got engaged, Bauer's parents held a celebratory reception; but attending it, Kafka confided to his diary, made him feel 'tied hand and foot like a criminal'. Shortly afterwards, during a rendezvous at a Berlin hotel, Kafka called off the engagement, but the letters continued. (Though Kafka was indecisive about those too: 'It is quite right that we should stop this business of so many letters,' he wrote to Bauer on one occasion, apparently in response to a suggestion of hers. 'Yesterday I even started a letter on this subject, which I will send tomorrow.') Two years later, the engagement was back on, but only for a while: in 1917, Kafka used the onset of tuberculosis as an excuse to cancel it a second and final time. It was presumably with some relief that Bauer married a banker, had two children, and moved to the United States, where she opened a successful knitwear firm – leaving behind her a liaison characterised by so many nightmarish and unpredictable reversals it's impossible to resist describing it as Kafkaesque.

It might be easy to file Kafka away under the heading of 'tortured genius', a remote figure with little relevance to our more ordinary lives. But the truth, as the critic Morris Dickstein writes, is that his 'neuroses are no different from ours, no more freakish: only more intense, more pure . . . [and] driven by genius to an integrity of unhappiness that most of us never approach'. Like the rest of us, Kafka railed at reality's constraints. He was indecisive in love, and in much else, because he yearned to live more than one life: to be a respectable citizen, which was why he kept his day job as an insurance claims investigator; to relate intimately to another person in marriage, which would mean marrying Bauer; and yet also to dedicate himself without compromise to his writing. On more than one occasion, in letters to Bauer, he characterised this struggle as a matter of 'two selves' wrestling with each other inside him – one in love with her but the other so consumed by literature that 'the death of his dearest friend would seem to be no more than a hindrance' to his work.

The degree of agony here might be extreme, but the essential tension is the same one felt by anybody torn between work and family, between a day job and a creative calling, a home town and the big city, or any other clash of possible lives. And Kafka responded like the rest of us, too, by trying not to confront the problem. Confining his relationship with Bauer to the realm of letters meant that he could cling to the possibility of a life of intimacy with her without allowing it to compete with his mania for work, as a real-life relationship necessarily would. This effort to dodge the implications of finitude doesn't always manifest itself in commitment-phobia like Kafka's: some people do commit outwardly to a relationship but hold back from full emotional commitment on the inside. Others find themselves years into threadbare marriages they actually should leave but don't, because they want to keep open the possibility that their relationship might yet blossom into a long and contented one, and also the option of exercising their freedom to leave at some future date. It's all the same essential evasion, though. At one point, a desperate-sounding Bauer advised her fiancé to try to 'live more in the real world'. But that was precisely what Kafka was seeking to avoid.

Six hundred miles away in Paris, and two decades before Franz met Felice, the French philosopher Henri Bergson tunnelled to the heart of Kafka's problem in his book *Time and Free Will.* We invariably prefer indecision over committing ourselves to a single path, Bergson wrote, because 'the future, which we dispose of to our liking, appears to us at the same time under a multitude of forms, equally attractive and equally possible'. In other words, it's easy for me to fantasise about, say, a life spent achieving stellar professional success, while also excelling as a parent and partner, while also dedicating myself to training for marathons or lengthy meditation retreats or volunteering in my community – because so long as I'm only fantasising, I get to imagine all of them unfolding simultaneously and flawlessly. As soon as I start trying to *live* any of those lives, though, I'll be forced to make trade-offs – to put less time than I'd like into one of those domains, so as to make space for another – and to accept that nothing I do will go perfectly anyway, with the result that my actual life will inevitably prove disappointing by comparison with the fantasy. 'The idea of the future, pregnant with an infinity of possibilities, is thus more fruitful than the future itself,' Bergson wrote, 'and this is why we find more charm in hope than in possession, in dreams than in reality.' Once again, the seemingly dispiriting message here is actually a liberating one. Since every real-world choice about how to live entails the loss of countless alternative ways of living, there's no reason to procrastinate, or to resist making commitments, in the anxious hope that you might somehow be able to avoid those losses. Loss is a given. That ship has sailed – and what a relief.

The Inevitability of Settling

Which brings me to one of the few pieces of dating advice I feel entirely confident in delivering, though in fact it's relevant in every other sphere of life, too. It concerns 'settling' – the ubiquitous modern fear that you might find yourself committing to a romantic partner who falls short of your ideal, or who's unworthy of your excellent personality. (The career-related version of this worry entails 'settling' for a job that pays the bills rather than going all-in on your passion.) The received wisdom, articulated in a thousand magazine articles and inspirational Instagram memes, is that it's always a crime to settle. But the received wisdom is wrong. You should definitely settle.

Or to be more precise, you don't have a choice. You *will* settle – and this fact ought to please you. The American political theorist Robert Goodin wrote a whole treatise on this topic, *On Settling*, in which he demonstrates, to start with, that we're inconsistent when it comes to what we define as 'settling'. Everyone seems to agree that if you embark on a relationship when you secretly suspect you could find someone better, you're guilty of settling, because you're opting to use up a portion of your life with a less-than-ideal partner. But since time is finite, the decision to *refuse* to settle – to spend a decade restlessly scouring online dating networks for the perfect person – is also a case of settling, because you're opting to use up a decade of your limited time in a different sort of less-than-ideal situation. Moreover, Goodin observes, we tend to contrast a life of settling with a life of what he labels 'striving', or living life to the fullest. But this is a mistake, too, and not just because settling is unavoidable but also because living life to the fullest *requires* settling. 'You must settle, in a relatively enduring way, upon something that will be the object of your striving, in order for that striving to count as striving,' he writes: you can't become an ultra-successful lawyer or artist or politician without first 'settling' on law, or art, or politics, and therefore deciding to forgo the potential rewards of other careers. If you flit between them all, you'll succeed in none of them. Likewise, there's no possibility of a romantic relationship being truly fulfilling unless you're willing, at least for a while, to settle for that specific relationship, with all its imperfections – which means spurning the seductive lure of an infinite number of superior imaginary alternatives.

Of course, we rarely approach relationships with such wisdom. Instead, we spend years failing to fully commit to any one relationship – either by finding a reason to call things off as soon as a serious liaison starts to look likely or by only half-heartedly showing up for whatever relationship we're in. Or, alternatively, in a pattern that every experienced psychotherapist has encountered a hundred times, we do commit – but then, after three or four years, start thinking about breaking things off, convinced that our partner's psychological issues are making things impossible, or that we're not as compatible as we'd believed. Either of these might conceivably be true in certain cases; people are sometimes guilty of spectacularly bad choices in love, and in other domains as well. But more often, the real problem is just that the other person is one other person. In other words, the cause of your difficulties isn't that your partner is especially flawed, or that the two of you are especially incompatible, but that you're finally noticing all the ways in which your partner is (inevitably) finite, and thus deeply disappointing by comparison with the world of your fantasy, where the limiting rules of reality don't apply.

The point that Bergson made about the future – that it's more appealing than the present because you get to indulge in all your hopes for it, even if they contradict each other – is no less true of fantasy romantic partners, who can easily exhibit a range of characteristics that simply couldn't coexist in one person in the real world. It's common, for example, to enter a relationship unconsciously hoping that your partner will provide both an unlimited sense of stability and an unlimited sense of excitement – and then, when that's not what transpires, to assume that the problem is your partner and that these qualities might coexist in someone else, whom you should therefore set off to find. The reality is that the demands are contradictory. The qualities that make someone a dependable source of excitement are generally the opposite of those that make him or her a dependable source of stability. Seeking both in one real human isn't much less absurd than dreaming of a partner who's both six and five feet tall.

And not only should you settle; ideally, you should settle in a way that makes it harder to back out, such as moving in together, or getting married, or having a child. The great irony of all our efforts to avoid facing finitude – to carry on believing that it might be possible not to have to choose between mutually exclusive options – is that when people finally *do* choose, in a relatively irreversible way, they're usually much happier as a result. We'll do almost anything to avoid burning our bridges, to keep alive the fantasy of a future unconstrained by limitation, yet having burned them, we're generally pleased that we did so. Once, in an experiment, the Harvard University social psychologist Daniel Gilbert and a colleague gave hundreds of people the opportunity to pick a free poster from a selection of art prints. Then he divided the participants into two groups. The first group was told that they had a month in which they could exchange their poster for any other one; the second group was told that the decision they'd already made had been final. In follow-up surveys, it was the latter group – those who were stuck with their decision, and who thus weren't distracted by the thought that it might still be possible to make a better choice – who showed by far the greater appreciation for the work of art they'd selected.

Not that we necessarily need psychologists to prove the point. Gilbert's study reflects an insight that's deeply embedded in numerous cultural traditions, most obviously that of marriage. When two spouses agree to stay together 'for better or worse', rather than bolting as soon as the going gets tough, they're making an agreement that not only will help them weather the rough patches, but that also promises to make the good times more fulfilling, too – because having committed themselves to one finite course of action, they'll be much less likely to spend that time pining after fantastical alternatives. In consciously making a commitment, they're closing off their fantasies of infinite possibility in favour of what I described, in the previous chapter, as the 'joy of missing out': the recognition that the renunciation of alternatives is what makes their choice a meaningful one in the first place. This is also why it can be so unexpectedly calming to take actions you'd been fearing or delaying – to finally hand in your notice at work, become a parent, address a festering family issue or exchange on a house. When you can no longer turn back, anxiety falls away, because now there's only one direction to travel: forward into the consequences of your choice.

5.

The Watermelon Problem

One Friday in April 2016, as that year's polarising American presidential race intensified, and more than thirty armed conflicts raged around the globe, approximately three million people spent part of their day watching two reporters from BuzzFeed wrap rubber bands around a watermelon. Gradually, over the course of forty-three agonising minutes, the pressure ramped up – both the psychological kind and the physical pressure on the watermelon – until, at minute forty-four, the 686th rubber band was applied. What happened next won't amaze you: the watermelon exploded, messily. The reporters high-fived, wiped the splatters from their reflective goggles, then ate some watermelon. The broadcast ended. The earth continued its orbit around the sun.

I'm not raising this to imply that there's anything especially shameful about spending forty-four minutes of your day staring at a watermelon on the internet. On the contrary, given what was to happen to life online in the years after 2016 – as the trolls and neo-Nazis began to crowd out the pop quizzes and cat videos, and social media increasingly became a matter of 'doomscrolling' in a depressive daze through bottomless feeds of bad news – the BuzzFeed watermelon escapade already feels like a tale from a happier time. But it's worth mentioning because it illustrates an elephant-in-the-room problem with everything I've been arguing so far about time and time management. That problem is distraction. After all, it hardly matters how committed you are to making the best use of your limited time if, day after day, your attention gets wrenched away by things on which you never wanted to focus. It's a safe bet that none of those three million people woke up that morning with the intention of using a portion of their lives to watch a watermelon burst; nor, when the moment arrived, did they necessarily feel as though they were freely *choosing* to do so. 'I want to stop watching so bad but I'm already committed,' read one typically rueful comment on Facebook. 'I've been watching you guys put rubber bands around a watermelon for 40 minutes,' wrote someone else. 'What am I doing with my life?'

The watermelon tale is a reminder, moreover, that these days distraction has become all but synonymous with digital distraction: it's what happens when the internet gets in the way of our attempts to concentrate. But this is misleading. Philosophers have been worrying about distraction at least since the time of the ancient Greeks, who saw it less as a matter of external interruptions and more as a question of character – a systematic inner failure to use one's time on what one claimed to value the most. Their reason for treating distraction so seriously was straightforward, and it's the reason we ought to do so, too: what you pay attention to will define, for you, what reality is.

Even commentators who spend a lot of time fretting about the modern-day 'crisis of distraction' rarely seem to grasp the full implications of this. For example, you hear it said that attention is a 'finite resource', and finite it certainly is: according to one calculation, by the psychologist Timothy Wilson, we're capable of consciously attending to about 0.0004 per cent of the information bombarding our brains at any given moment. But to describe attention as a 'resource' is to subtly misconstrue its centrality in our lives. Most other resources on which we rely as individuals – such as food, money and electricity – are things that *facilitate* life, and in some cases it's possible to live without them, at least for a while. Attention, on the other hand, just *is* life: your experience of being alive consists of nothing other than the sum of everything to which you pay attention. At the end of your life, looking back, whatever compelled your attention from moment to moment is simply what your life will have been. So when you pay attention to something you don't especially value, it's not an exaggeration to say that you're paying with your life. Seen this way, 'distraction' needn't refer only to momentary lapses in focus, as when you're distracted from performing your work duties by the ping of an incoming text message, or a compellingly terrible news story. The job itself could be a distraction – that is, an investment of a portion of your attention, and therefore of your life, in something less meaningful than other options that might have been available to you.

This was why Seneca, in *On the Shortness of Life*, came down so hard on his fellow Romans for pursuing political careers they didn't really care about, holding elaborate banquets they didn't especially enjoy, or just 'baking their bodies in the sun': they didn't seem to realise that in succumbing to such diversions, they were squandering the very stuff of existence. Seneca risks sounding like an uptight pleasure-hater here – after all, what's so bad about a bit of sunbathing? – and to be honest, I suspect he probably was. But the crucial point isn't that it's wrong to choose to spend your time relaxing, whether at the beach or on BuzzFeed. It's that the distracted person isn't really choosing at all. Their attention has been commandeered by forces that don't have their highest interests at heart.

The proper response to this situation, we're often told today, is to render ourselves indistractible in the face of interruptions: to learn the secrets of 'relentless focus' – usually involving meditation, web-blocking apps, expensive noise-cancelling headphones, and more meditation – so as to win the attentional struggle once and for all. But this is a trap. When you aim for this degree of control over your attention, you're making the mistake of addressing one truth about human limitation – your limited time, and the consequent need to use it well – by denying another truth about human limitation, which is that achieving total sovereignty over your attention is almost certainly impossible. In any case, it would be highly undesirable to be able to do exactly as you wished with your attention. If outside forces couldn't commandeer at least some of it against your will, you'd be unable to step out of the path of oncoming buses, or hear that your baby was in distress. Nor are the benefits confined to emergencies; the same phenomenon is what allows your attention to be seized by a beautiful sunset, or your eye to be caught by a stranger's across a room. But it's the obvious survival advantages of this kind of distractibility that explain why we evolved that way. The Palaeolithic hunter-gatherer whose attention was alerted by a rustling in the bushes, whether he liked it or not, would have been far more likely to flourish than one who heard such rustlings only after first making the conscious decision to listen out for them.

Neuroscientists call this 'bottom-up' or involuntary attention, and we'd struggle to stay alive without it. Yet the capacity to exert some influence over the other part of your attention – the 'top-down' or voluntary kind – can make the whole difference between a well-lived life and a hellish one. The classic and extreme demonstration of this is the case of the Austrian psychotherapist Viktor Frankl, author of *Man's Search for Meaning*, who was able to fend off despair as a prisoner in Auschwitz because he retained the ability to direct a portion of his attention towards the only domain the camp guards couldn't violate: his inner life, which he was then able to conduct with a measure of autonomy, resisting the outer pressures that threatened to reduce him to the status of an animal. But the flip side of this inspiring truth is that a life spent in circumstances immeasurably better than a concentration camp can still end up feeling fairly meaningless if you're incapable of directing some of your attention as you'd like. After all, to have any meaningful experience, you must be able to focus on it, at least a bit. Otherwise, are you really *having* it at all? Can you have an experience you don't experience? The finest meal at a Michelin-starred restaurant might as well be a plate of instant noodles if your mind is elsewhere; and a friendship to which you never actually give a moment's thought is a friendship in name only. 'Attention is the beginning of devotion,' writes the poet Mary Oliver, pointing to the fact that distraction and care are incompatible with each other: you can't truly love a partner or a child, dedicate yourself to a career or to a cause – or just savour the pleasure of a stroll in the park – except to the extent that you can hold your attention on the object of your devotion to begin with.

A Machine for Misusing Your Life

All of which helps clarify what's so alarming about the contemporary online 'attention economy', of which we've heard so much in recent years: it's essentially a giant machine for persuading you to make the wrong choices about what to do with your attention, and therefore with your finite life, by getting you to care about things you didn't want to care about. And you have far too little control over your attention simply to decide, as if by fiat, that you're not going to succumb to its temptations.

Many of us are familiar by now with the basic contours of this situation. We know that the 'free' social media platforms we use aren't really free, because, as the saying goes, you're not the customer but the product being sold: in other words, the technology companies' profits come from seizing our attention, then selling it to advertisers. We're at least dimly aware, too, that our smartphones are tracking our every move, recording how we swipe and click, what we linger on or scroll past, so that the data collected can be used to show us precisely that content most likely to keep us hooked, which usually means whatever makes us angriest or most horrified. All the feuds and fake news and public shamings on social media, therefore, aren't a flaw, from the perspective of the platform owners; they're an integral part of the business model.

You might also be aware that all this is delivered by means of 'persuasive design' – an umbrella term for an armoury of psychological techniques borrowed directly from the designers of casino slot machines, for the express purpose of encouraging compulsive behaviour. One example among hundreds is the ubiquitous drag-down-to-refresh gesture, which keeps people scrolling by exploiting a phenomenon known as 'variable rewards': when you can't predict whether or not refreshing the screen will bring new posts to read, the uncertainty makes you more likely to keep trying, again and again and again, just as you would on a slot machine. When this whole system reaches a certain level of pitiless efficiency, the former Facebook investor turned detractor Roger McNamee has argued, the old cliché about users as 'the product being sold' stops seeming so apt. After all, companies are generally motivated to treat even their products with a modicum of respect, which is more than can be said about how some of them treat their users. A better analogy, McNamee suggests, is that we're the fuel: logs thrown on Silicon Valley's fire, impersonal repositories of attention to be exploited without mercy, until we're all used up.

What's far less widely appreciated than all that, though, is how deep the distraction goes, and how radically it undermines our efforts to spend our finite time as we'd like. As you surface from an hour inadvertently frittered away on Facebook, you'd be forgiven for assuming that the damage, in terms of wasted time, was limited to that single misspent hour. But you'd be wrong. Because the attention economy is designed to prioritise whatever's most compelling – instead of whatever's most true, or most useful – it systematically distorts the picture of the world we carry in our heads at all times. It influences our sense of what matters, what kinds of threats we face, how venal our political opponents are, and thousands of other things – and all these distorted judgements then influence how we allocate our offline time as well. If social media convinces you, for example, that violent crime is a far bigger problem in your city than it really is, you might find yourself walking the streets with unwarranted fear, staying home instead of venturing out, and avoiding interactions with strangers – and voting for a demagogue with a tough-on-crime platform. If all you ever see of your ideological opponents online is their very worst behaviour, you're liable to assume that even family members who differ from you politically must be similarly, irredeemably bad, making relationships with them hard to maintain. So it's not simply that our devices distract us from more important matters. It's that they change how we're defining 'important matters' in the first place. In the words of the philosopher

Harry Frankfurt, they sabotage our capacity to 'want what we want to want'.

My own squalid, but I suspect entirely typical, history as a Twitter junkie might serve as a case in point. Even at the height of my dependency (I'm now in recovery), I rarely spent more than two hours a day glued to the screen. Yet Twitter's dominion over my attention extended a great deal further than that. Long after I'd closed the app, I'd be panting on the treadmill at the gym, or chopping carrots for dinner, only to find myself mentally prosecuting a devastating argument against some idiotic holder of Wrong Opinions I'd had the misfortune to encounter online earlier that day. (It wasn't misfortune really, of course; the algorithm showed me those posts deliberately, having learned what would wind me up.) Or my newborn son would do something adorable, and I'd catch myself speculating about how I might describe it in a tweet, as if what mattered wasn't the experience but my (unpaid!) role as a provider of content for Twitter. And I vividly recall walking alone along a windswept Scottish beach, as dusk began to fall, when I experienced one particularly disturbing side effect of 'persuasive design', which is the twitchiness you start to feel when the activity in which you're engaged *hasn't* been crafted by a team of professional psychologists hell-bent on ensuring that your attention never wavers. I love windswept Scottish beaches at dusk more passionately than anything I can ever remember encountering on social media. But only the latter is engineered to constantly adapt to my interests and push my psychological buttons, so as to keep my attention captive. No wonder the rest of reality sometimes seems unable to compete.

At the same time, the hopelessness of the world I encountered online began to seep into the world of the concrete. It was impossible to drink from Twitter's fire hose of anger and suffering – of news and opinions selected for my perusal precisely because they weren't the norm, which was what made them especially compelling – without starting to approach the rest of life as if they *were* the norm, which meant being constantly braced for confrontation or disaster, or harbouring a nebulous sense of foreboding. Unsurprisingly, this rarely proved to be the basis for a fulfilling day. To make things more troublesome still, it can be difficult even to notice when your outlook on life is being changed in this depressing fashion, thanks to a special problem with attention, which is that it's extremely difficult for it to monitor itself. The only faculty you can use to see what's happening to your attention is your attention, the very thing that's already been commandeered. This means that once the attention economy has rendered you sufficiently distracted, or annoyed, or on edge, it becomes easy to assume that this is just what life these days inevitably feels like. In T. S. Eliot's words, we are 'distracted from distraction by distraction'. The unsettling possibility is that if you're convinced that none of this is a problem for you – that social media *hasn't* turned you into an angrier, less empathetic, more anxious or more numbed-out version of yourself – that might be because it has. Your finite time has been appropriated, without your realising anything's amiss.

It's been obvious for some time now, of course, that all this constitutes a political emergency. By portraying our opponents as beyond persuasion, social media sorts us into ever more hostile tribes, then rewards us, with likes and shares, for the most hyperbolic denunciations of the other side, fuelling a vicious cycle that makes sane debate impossible. Meanwhile, we've learned the hard way that unscrupulous politicians can overwhelm their opposition, not to mention the fact-checking capabilities of journalists, simply by flooding a nation's attentional bandwidth with outrage after outrage, so that each new scandal overwrites the last one in public awareness – and anyone who responds or retweets, even if their intention is to condemn the hatemongering, finds themselves rewarding it with attention, thereby helping it spread.

As the technology critic Tristan Harris likes to say, each time you open a social media app, there are 'a thousand people on the other side of the screen' paid to keep you there – and so it's unrealistic to expect users to resist the assault on their time and attention by means of willpower alone. Political crises demand political solutions. Yet if we're to understand distraction at the deepest level, we'll also have to acknowledge an awkward truth at the bottom of all this, which is that 'assault' – with its implications of an uninvited attack – isn't quite the right word. We mustn't let Silicon Valley off the hook, but we should be honest: much of the time, we give in to distraction willingly. Something in us wants to be distracted, whether by our digital devices or anything else – to *not* spend our lives on what we thought we cared about the most. The calls are coming from inside the house. This is among the most insidious of the obstacles we face in our efforts to use our finite lives well, so it's time to take a closer look at it.

6.

The Intimate Interrupter

Had you been walking in the Kii Mountains in southern Japan during the winter months of 1969, you might have witnessed something startling: a pale and skinny American man, entirely naked, dumping half-frozen water over his own head from a large wooden cistern. His name was Steve Young, and he was training to become a monk in the Shingon branch of Buddhism – but so far the process had been nothing but a sequence of humiliations. First, the abbot of the Mount Koya monastery had refused to let him in the door. Who on earth was this gangly white Asian studies PhD student, who'd apparently decided the life of a Japanese monk was for him? Eventually, after some badgering, Young had been permitted to stay, but only in return for performing various menial tasks around the monastery, like sweeping the hallways and washing dishes. Now, at last, he had been authorised to begin the hundred-day solo retreat that marked the first real step on the monastic journey – only to discover that it entailed living in a tiny unheated hut and conducting a thrice-daily purification ritual in which Young, who'd been raised beside the ocean in balmy California, had to douse himself with several gallons of bone-chilling melted snow. It was a 'horrific ordeal', he would recall years later. 'It's so cold that the water freezes the moment it touches the floor, and your towel freezes in your hand. So you're sliding around barefoot on ice, trying to dry your body with a frozen hand towel.'

Faced with physical distress – even of a much milder variety than this – most people's instinctive reaction is to try not to pay attention to it, to attempt to focus on anything else at all. For example, if you're mildly phobic about hypodermic syringes, like I am, you've probably found yourself staring very hard at the mediocre artwork in doctors' clinics in an effort to take your mind off the jab you're about to receive. At first, this was Young's instinct, too: to recoil internally from the experience of the freezing water hitting his skin by thinking about something different – or else just trying, through an act of sheer will, not to feel the cold. This is hardly an unreasonable reaction: when it's so unpleasant to stay focused on present experience, common sense would seem to suggest that mentally absenting yourself from the situation would moderate the pain.

And yet as icy deluge followed icy deluge, Young began to understand that this was precisely the wrong strategy. In fact, the more he concentrated on the sensations of intense cold, giving his attention over to them as completely as he could, the less agonising he found them – whereas once his 'attention wandered, the suffering became unbearable'. After a few days, he began preparing for each drenching by first becoming as focused on his present experience as he possibly could so that, when the water hit, he would avoid spiralling from mere discomfort into agony. Slowly it dawned on him that this was the whole point of the ceremony. As he put it – though traditional Buddhist monks certainly would not have done so – it was a 'giant biofeedback device', designed to train him to concentrate by rewarding him (with a reduction in suffering) for as long as he could remain undistracted, and punishing him (with an increase in suffering) whenever he failed. After his retreat, Young – who is now a meditation teacher better known as Shinzen Young, his new first name having been bestowed on him by the abbot at Mount Koya – found that his powers of concentration had been transformed. Whereas staying focused on the present had made the agonies of the ice-water ritual more tolerable, it made less unpleasant undertakings – daily chores that might previously have been a source not of agony but of boredom or annoyance – positively engrossing. The more intensely he could hold his attention on the experience of whatever he was doing, the clearer it became to him that the real problem had been not the activity itself but his internal resistance to experiencing it. When he stopped trying to block out those sensations and attended to them instead, the discomfort would evaporate.

Young's ordeal demonstrates an important point about what's going on when we succumb to distraction, which is that we're motivated by the desire to try to flee something painful about our experience of the present. This is obvious enough when the pain in question is physical, like icy water on naked skin and a flu jab at the doctor's surgery – cases in which the difficult sensations are so hard to ignore that it takes real effort to shift your attention elsewhere. But it's also true, in a subtler way, when it comes to everyday distraction. Consider the archetypal case of being lured from your work by social media: it's not usually that you're sitting there, concentrating rapturously, when your attention is dragged away against your will. In truth, you're eager for the slightest excuse to turn away from what you're doing, in order to escape how disagreeable it feels to be doing it; you slide away to the Twitter pile-on or the celebrity gossip site with a feeling not of reluctance but of relief. We're told that there's a 'war for our attention', with Silicon Valley as the invading force. But if that's true, our role on the battlefield is often that of collaborators with the enemy.

Mary Oliver calls this inner urge towards distraction 'the intimate interrupter' – that 'self within the self, that whistles and pounds upon the door panels', promising an easier life if only you'd redirect your attention away from the meaningful but challenging task at hand, to whatever's unfolding one browser tab away. 'One of the puzzling lessons I have learned,' observes the American author Gregg Krech, describing his own experience of the same urge, 'is that, more often than not, I do not feel like doing most of the things that need doing. I'm not just speaking about cleaning the toilet bowl or doing my tax returns. I'm referring to those things I genuinely desire to accomplish.'

The Discomfort of What Matters

It's worth pausing to notice how exceptionally strange this is. Why, exactly, are we rendered so uncomfortable by concentrating on things that matter – the things we thought we wanted to do with our lives – that we'd rather flee into distractions, which, by definition, are what we *don't* want to be doing with our lives? Certain specific tasks might be so unpleasant or intimidating that a preference for avoiding them wouldn't be very remarkable. But the more common issue is one of boredom, which often arises without explanation. Suddenly, the thing you'd resolved to do, because it mattered to you to do it, feels so staggeringly tedious that you can't bear to focus on it for one moment more.

The solution to this mystery, dramatic though it might sound, is that whenever we succumb to distraction, we're attempting to flee a painful encounter with our finitude – with the human predicament of having limited time, and more especially, in the case of distraction, limited control over that time, which makes it impossible to feel certain about how things will turn out. (Except, that is, for the deeply unpleasant certainty that one day death will bring it all to an end.) When you try to focus on something you deem important, you're forced to face your limits, an experience that feels especially uncomfortable precisely because the task at hand is one you value so much. Unlike the architect from Shiraz, who refused to bring his ideal mosque into the world of time and imperfection, you're obliged to give up your godlike fantasies and to experience your lack of power over things you care about. Perhaps the cherished creative project will prove beyond your talents, or maybe the difficult marital conversation for which you'd been steeling yourself will unravel into a bitter argument. And even if everything proceeds wonderfully, you couldn't have known in advance that it was going to do so, so you'll still have had to give up the feeling of being the master of your time. To quote the psychotherapist Bruce Tift once more, you'll have had to allow yourself to risk feeling 'claustrophobic, imprisoned, powerless, and constrained by reality'.

This is why boredom can feel so surprisingly, aggressively unpleasant: we tend to think of it merely as a matter of not being particularly interested in whatever it is we're doing, but in fact it's an intense reaction to the deeply uncomfortable experience of confronting your limited control. Boredom can strike in widely differing contexts – when you're working on a major project; when you can't think of anything to do on a Sunday afternoon; when it's your job to care for a two-year-old for five hours straight – but they all have one characteristic in common: they demand that you face your finitude. You're obliged to deal with how your experience is unfolding in this moment, to resign yourself to the reality that *this is it*.

No wonder we seek out distractions online, where it feels as though no limits apply – where you can update yourself instantaneously on events taking place a continent away, present yourself however you like, and keep scrolling forever through infinite newsfeeds, drifting through 'a realm in which space doesn't matter and time spreads out into an endless present', to quote the critic James Duesterberg. It's true that killing time on the internet often doesn't feel especially *fun*, these days. But it doesn't need to feel fun. In order to dull the pain of finitude, it just needs to make you feel unconstrained.

This also makes it easier to see why the strategies generally recommended for defeating distraction – digital detoxes, personal rules about when you'll allow yourself to check your inbox, and so forth – rarely work, or at least not for long. They involve limiting your access to the things you use to assuage your urge towards distraction, and in the case of the most addictive forms of technology, that's surely a sensible idea. But they don't address the urge itself. Even if you quit Facebook, or ban yourself from social media during the workday, or exile yourself to a cabin in the mountains, you'll probably still find it unpleasantly constraining to focus on what matters, so you'll find some way to relieve the pain by distracting yourself: by daydreaming, taking an unnecessary nap, or – the preferred option of the productivity geek – redesigning your to-do list and reorganising your desk.

The overarching point is that what we think of as 'distractions' aren't the ultimate cause of our being distracted. They're just the places we go to seek relief from the discomfort of confronting limitation. The reason it's hard to focus on a conversation with your spouse isn't that you're surreptitiously checking your phone beneath the dinner table. On the contrary, 'surreptitiously checking your phone beneath the dinner table' is what you do *because* it's hard to focus on the conversation – because listening takes effort and patience and a spirit of surrender, and because what you hear might upset you, so checking your phone is naturally more pleasant. Even if you place your phone out of reach, therefore, you shouldn't be surprised to find yourself seeking some other way to avoid paying attention. In the case of conversation, this generally takes the form of mentally rehearsing what you're going to say next, as soon as the other person has finished making sounds with their mouth.

I wish I could reveal, at this point, the secret for uprooting the urge towards distraction – the way to have it not feel unpleasant to decide to hold your attention, for a sustained time, on something you value, or that you can't easily choose not to do. But the truth is that I don't think there is one. The most effective way to sap distraction of its power is just to stop expecting things to be otherwise – to accept that this unpleasantness is simply what it feels like for finite humans to commit ourselves to the kinds of demanding and valuable tasks that force us to confront our limited control over how our lives unfold.

Yet there's a sense in which accepting this lack of any solution *is* the solution. Young's discovery on the mountainside, after all, was that his suffering subsided only when he resigned himself to the truth of his situation: when he stopped fighting the facts and allowed himself to more fully feel the icy water on his skin. The less attention he devoted to objecting to what was happening to him, the more attention he could give to what was actually happening. My powers of concentration might not come close to Young's, but I've found the same logic applies. The way to find peaceful absorption in a difficult project, or a boring Sunday afternoon, isn't to chase feelings of peace or absorption, but to acknowledge the inevitability of discomfort, and to turn more of your attention to the reality of your situation than to railing against it.

Some Zen Buddhists hold that the entirety of human suffering can be boiled down to this effort to resist paying full attention to the way things are going, because we wish they were going differently ('This shouldn't be happening!'), or because we wish we felt more in control of the process. There is a very down-to-earth kind of liberation in grasping that there are certain truths about being a limited human from which you'll never be liberated. You don't get to dictate the course of events. And the paradoxical reward for accepting reality's constraints is that they no longer feel so constraining.

Part II

Beyond Control

7.

We Never Really Have Time

The cognitive scientist Douglas Hofstadter is famous, among other reasons, for coining 'Hofstadter's law', which states that any task you're planning to tackle will always take longer than you expect, 'even when you take into account Hofstadter's Law'. In other words, even if you know that a given project is likely to overrun, and you adjust your schedule accordingly, it'll just overrun your new estimated finishing time, too. It follows from this that the standard advice about planning – to give yourself twice as long as you think you'll need – could actually make matters worse. You might be well aware of, say, your unrealistic tendency to assume that you can complete the weekly grocery shopping in an hour, door to door. But if you allow yourself two hours, precisely because you *know* that you're usually over-optimistic, you may find it taking two and a half hours instead. (The effect becomes especially clear on a bigger scale: the government of New South Wales, being acutely conscious that big construction projects tend to overrun, allowed a seemingly ample four years for the building of the Sydney Opera House – but it ended up taking fourteen, at a cost of more than 1,400 per cent of the original budget.) Hofstadter was half joking, of course. But I've always found something a little unsettling about his law, because if it's true – and it certainly seems to be, in my experience – it suggests something very strange: that the activities we try to plan for somehow actively resist our efforts to make them conform to our plans. It's as if our efforts to be good planners don't merely fail but cause things to take longer still. Reality seems to fight back, an angry god determined to remind us that it retains the upper hand, no matter how much we try to supplicate to it by building extra slack into our schedules.

To be fair, this sort of thing probably bothers me more than most, because I come from a family of people you might reasonably call obsessive planners. We're the type who like to get our ducks in a row by confirming, as far in advance as possible, how the future is going to unfold, and who get antsy and anxious when obliged to coordinate with those who prefer to take life as it comes. My wife and I are lucky to make it to the end of June, in any given year, before receiving the first enquiry from my parents about our plans for Christmas; and I was raised to regard anyone who booked a flight or hotel room less than about four months before the proposed date of departure or occupancy as living life on the edge to an inexcusable degree. On family holidays, we could be guaranteed a three-hour wait at the airport, or an hour at the railway station, having left home much too far ahead of time. ('Dad Suggests Arriving at Airport 14 Hours Early,' reads a headline in *The Onion*, apparently inspired by my childhood.) All this annoyed me then, as it annoys me today, with that special irritation reserved for traits one recognises all too clearly in oneself as well.

At least I think I can say that my family comes by it honestly. My paternal grandmother, who was Jewish, was nine years old and living in Berlin when Hitler came to power in 1933, and she was fifteen by the time her stepfather, surveying the wreckage of Kristallnacht, finally made plans to get his family to Hamburg and thence on board the SS *Manhattan*, bound for Southampton in England. (The passengers, I was once told, popped champagne corks on deck, but only after they were certain the ship had left German waters.) Her own grandmother, my great-great-grandmother, never made it out, and later died in the concentration camp at Theresienstadt. It's not especially hard to see how an adolescent German-Jewish girl, arriving in London on the eve of the Second World War, might acquire, and later pass on to her children, the unshakeable belief that if you didn't plan things exactly right, some very bad fate might befall you or those you loved. Sometimes, when you're leaving on a trip, it really is important to get to your point of departure in plenty of time.

The trouble with being so emotionally invested in planning for the future, though, is that while it may occasionally prevent a catastrophe, the rest of the time it tends to exacerbate the very anxiety it was supposed to allay. The obsessive planner, essentially, is demanding certain reassurances from the future – but the future isn't the sort of thing that can ever provide the reassurance he craves, for the obvious reason that it's still in the future. After all, you can never be absolutely certain that something won't make you late for the airport, no matter how many spare hours you build in. Or rather you can be certain – but only once you've arrived and you're cooling your heels in the terminal, at which point there's no solace to be gained from the fact that everything turned out fine, because that's all in the past now, and there's the next chunk of the future to feel anxious about instead. (Will the plane land at its destination in time for you to catch your onward train? And so on and so on.) Really, no matter how far ahead you plan, you never get to relax in the certainty that everything's going to go the way you'd like. Instead, the frontier of your uncertainty just gets pushed further and further towards the horizon. Once your Christmas plans are nailed down, there's January to think about, then February, then March . . .

I'm using my neurotic family by way of example here, but it's important to see that this underlying longing to turn the future into something dependable isn't confined to compulsive planners. It's present in anyone who worries about anything, whether or not they respond by devising elaborate timetables or hyper-cautious travel plans. Worry, at its core, is the repetitious experience of a mind attempting to generate a feeling of security about the future, failing, then trying again and again and again – as if the very effort of worrying might somehow help forestall disaster. The fuel behind worry, in other words, is the internal demand to know, in advance, that things will turn out fine: that your partner *won't* leave you, that you *will* have sufficient money to retire, that a pandemic *won't* claim the lives of anyone you love, that your favoured candidate *will* win the next election, that you *can* get through your to-do list by the end of Friday afternoon. But the struggle for control over the future is a stark example of our refusal to acknowledge our built-in limitations when it comes to time, because it's a fight the worrier obviously won't win. You can never be truly certain about the future. And so your reach will always exceed your grasp.

Anything Could Happen

In much of this book so far, I've emphasised the importance of confronting, rather than avoiding, the uncomfortable reality about how little time we have. But it should also be becoming clear that there's something suspect about the idea of time as a thing we 'have' in the first place. As the writer David Cain points out, we never have time in the same sense that we have the cash in our wallets or the shoes on our feet. When we claim that we have time, what we really mean is that we expect it. 'We assume we have three hours or three days to do something,' Cain writes, 'but it never actually comes into our possession.' Any number of factors could confound your expectations, robbing you of the three hours you thought you 'had' in which to complete an important work project: your boss could interrupt with an urgent request; the Tube could break down; you could die. And even if you do end up getting the full three hours, precisely in line with your expectations, you won't know this for sure until the point at which those hours have passed into history. You only ever get to feel certain about the future once it's already turned into the past.

Likewise, and despite everything I've been saying, nobody ever really gets four thousand weeks in which to live – not only because you might end up with fewer than that, but because in reality you never even *get* a single week, in the sense of being able to guarantee that it will arrive, or that you'll be in a position to use it precisely as you wish. Instead, you just find yourself in each moment as it comes, already thrown into this time and place, with all the limitations that entails, and unable to feel certain about what might happen next. Reflect on this a little, and Heidegger's idea that we are time – that there's no meaningful way to think of a person's existence except as a sequence of moments of time – begins to make more sense. And it has real psychological consequences, because the assumption that time is something we can possess or control is the unspoken premise of almost all our thinking about the future, our planning and goal-setting and worrying. So it's a constant source of anxiety and agitation, because our expectations are forever running up against the stubborn reality that time isn't in our possession and can't be brought under our control.

My point, to be clear, isn't that it's a bad idea to make plans, or save money for retirement, or remember to vote, so as to increase the chances that the future will turn out the way you'd like. Our efforts to influence the future aren't the problem. The problem – the source of all the anxiety – is the need that we feel, from our vantage point here in the present moment, to be able to know that those efforts will prove successful. It's fine, of course, to strongly prefer that your partner never leave you, and to treat him or her in ways that make that happy outcome more likely. But it's a recipe for a life of unending stress to insist that you must be able to feel certain, now, that this is how your relationship is definitely going to unfold in the future. So a surprisingly effective antidote to anxiety can be to simply realise that this demand for reassurance from the future is one that will definitely never be satisfied – no matter how much you plan or fret, or how much extra time you leave to get to the airport. You *can't* know that things will turn out all right. The struggle for certainty is an intrinsically hopeless one – which means you have permission to stop engaging in it. The future just isn't the sort of thing you get to order around like that, as the French mathematician and philosopher Blaise Pascal understood: 'So imprudent are we,' he wrote, 'that we wander in the times which are not ours . . . We try to [give the present the support of] the future, and think of arranging matters which are not in our power, for a time which we have no certainty of reaching.'

Our anxiety about the uncontrollability of the future begins to seem rather more absurd, and perhaps therefore a little easier to let go of, when considered in the context of the past. We go through our days fretting because we can't control what the future holds; and yet most of us would probably concede that we got to wherever we are in our lives without exerting much control over it at all. Whatever you value most about your life can always be traced back to some jumble of chance occurrences you couldn't possibly have planned for, and that you certainly can't alter retrospectively now. You might never have been invited to the party where you met your future spouse. Your parents might never have moved to the neighbourhood near the school with the inspiring teacher who perceived your undeveloped talents and helped you shine. And so on – and if you peer back even further in time, to before your own birth, it's an even more dizzying matter of coincidence piled upon coincidence. In her autobiography *All Said and Done*, Simone de Beauvoir marvels at the mind-boggling number of things, all utterly beyond her control, that had to happen in order to make her *her*:

> If I go to sleep after lunch in the room where I work, sometimes I wake up with a feeling of childish amazement – why am I myself? What astonishes me, just as it astonishes a child when he becomes aware of his own identity, is the fact of finding myself here, and at this moment, deep in this life and not in any other. What stroke of chance has brought this about? . . . The penetration of that particular ovum by that particular spermatozoon, with its implications of the meeting of my parents and before that of their birth and the birth of all their forebears, had not one chance in hundreds of millions of coming about. And it is chance, a chance quite unpredictable in the present state of science, that caused me to be born a woman. From that point on, it seems to me that a thousand different futures might have stemmed from every single movement of my past: I might have fallen ill and broken off my studies; I might not have met Sartre; anything at all might have happened.

There's a soothing implication in de Beauvoir's words: that despite our total lack of control over any of these occurrences, each of us made it through to this point in our lives – so it might at least be worth entertaining the possibility that when the uncontrollable future arrives, we'll have what it takes to weather that as well. And that you shouldn't necessarily even want such control, given how much of what you value in life only ever came to pass thanks to circumstances you never chose.

Minding Your Own Business

These truths about the uncontrollability of the past and the unknowability of the future explain why so many spiritual traditions seem to converge on the same advice: that we should aspire to confine our attentions to the only portion of time that really *is* any of our business – this one, here in the present. 'Trying to control the future is like trying to take the master carpenter's place,' cautions one of the founding texts of Taoism, the Tao Te Ching, in a warning echoed several centuries later by the Buddhist scholar Geshe Shawopa, who gruffly commanded his students, 'Do not rule over imaginary kingdoms of endlessly proliferating possibilities.' Jesus says much the same thing in the Sermon on the Mount (though many of his later followers would interpret the Christian idea of eternal life as a reason to fixate on the future, not to ignore it). 'Take no thought for the morrow, for the morrow shall take thought for the things of itself,' he advises. Then he adds the celebrated phrase 'sufficient to the day is the evil thereof', a line I've only ever been able to hear in a tone of wry amusement directed at his listeners: Do you first-century working-class Galileans really lead such problem-free lives, he seems to be teasing them, that it makes sense to invent additional problems by fretting about what might happen tomorrow?

But the version of this thought that has always resonated the most for me comes from the modern-day spiritual teacher Jiddu Krishnamurti, who expressed it, in a characteristically direct manner, in a lecture delivered in California in the late 1970s. 'Partway through this particular talk,' recalls the writer Jim Dreaver, who was in attendance, 'Krishnamurti suddenly paused, leaned forward, and said, almost conspiratorially, "Do you want to know what my secret is?" Almost as though we were one body, we sat up . . . I could see people all around me lean forward, their ears straining, their mouths slowly opening in hushed anticipation.' Then Krishnamurti 'said in a soft, almost shy voice, "You see, I don't mind what happens."'

I don't mind what happens. Perhaps these words need a little unpacking; I don't think Krishnamurti means to say that we shouldn't feel sorrow, compassion or anger when bad things happen to ourselves or others, nor that we should give up on our efforts to prevent bad things from happening in the future. Rather, a life spent 'not minding what happens' is one lived without the inner demand to know that the future will conform to your desires for it – and thus without having to be constantly on edge as you wait to discover whether or not things will unfold as expected. None of that means we can't act wisely in the present to reduce the chances of bad developments later on. And we can still respond, to the best of our abilities, should bad things nonetheless occur; we're not obliged to accept suffering or injustice as part of the inevitable order of things. But to the extent that we can stop demanding certainty that things will go our way later on, we'll be liberated from anxiety in the only moment it ever actually is, which is this one.

Incidentally, I also don't take Krishnamurti to be recommending that we emulate those irritating individuals (we all know one or two of them) who are a little *too* proud of their commitment to being spontaneous – who insist on their right to never make plans and to skip impulsively through life, and of whom you can never be sure that an agreement to meet at six o'clock for a drink means they have the slightest intention of showing up. These ostentatiously free-and-easy types seem to feel confined by the very act of making plans, or trying to stick to them. But planning is an essential tool for constructing a meaningful life, and for exercising our responsibilities towards other people. The real problem isn't planning. It's that we take our plans to be something they aren't. What we forget, or can't bear to confront, is that, in the words of the American meditation teacher Joseph Goldstein, 'a plan is just a thought'. We treat our plans as though they are a lasso, thrown from the present around the future, in order to bring it under our command. But all a plan is – all it could ever possibly be – is a present-moment statement of intent. It's an expression of your current thoughts about how you'd ideally like to deploy your modest influence over the future. The future, of course, is under no obligation to comply.

8.

You Are Here

There's another sense in which treating time as something that we own and get to control seems to make life worse. Inevitably, we become obsessed with 'using it well', whereupon we discover an unfortunate truth: the more you focus on using time well, the more each day begins to feel like something you have to *get through*, en route to some calmer, better, more fulfilling point in the future, which never actually arrives. The problem is one of instrumentalisation. To use time, by definition, is to treat it instrumentally, as a means to an end, and of course we do this every day: you don't boil the kettle out of a love of boiling kettles, or put your socks in the washing machine out of a love for operating washing machines, but because you want a cup of coffee or clean socks. Yet it turns out to be perilously easy to over-invest in this instrumental relationship to time – to focus exclusively on where you're headed, at the expense of focusing on where you are – with the result that you find yourself living mentally in the future, locating the 'real' value of your life at some time that you haven't yet reached, and never will.

In his book *Back to Sanity*, the psychologist Steve Taylor recalls watching tourists at the British Museum in London who weren't really looking at the Rosetta Stone, the ancient Egyptian artefact on display in front of them, so much as preparing to look at it later, by recording images and videos of it on their phones. So intently were they focused on using their time for a future benefit – for the ability to revisit or share the experience later on – that they were barely experiencing the exhibition itself at all. (And who ever watches most of those videos anyway?) Of course, grumbling about younger people's smartphone habits is a favourite pastime of middle-aged curmudgeons like Taylor and me. But his deeper point is that we're all frequently guilty of something similar. We treat everything we're doing – life itself, in other words – as valuable only insofar as it lays the groundwork for something else.

This future-focused attitude often takes the form of what I once heard described as the '"when-I-finally" mind', as in: 'When I finally get my workload under control/get my candidate elected/find the right romantic partner/sort out my psychological issues, *then* I can relax, and the life I was always meant to be living can begin.' The person mired in this mentality believes that the reason she doesn't feel fulfilled and happy is that she hasn't yet managed to accomplish certain specific things; when she does so, she imagines, she'll feel in charge of her life and be the master of her time. Yet in fact the way she's attempting to achieve that sense of security means she'll *never* feel fulfilled, because she's treating the present solely as a path to some superior future state – and so the present moment won't ever feel satisfying in itself. Even if she does get her workload under control, or meet her soulmate, she'll just find some other reason to postpone her fulfilment until later on.

Certainly, context matters; there are plenty of situations in which it's understandable that people focus intently on the possibility of a better future. Nobody faults the low-paid cleaner of public toilets for looking forward to the end of the day, or to a time in the future when he has a better job; in the meantime, he naturally treats his working hours mainly as a means to the end of receiving a pay cheque. But there's something odder about the ambitious and well-paid architect, employed in the profession she always longed to join, who nonetheless finds herself treating every moment of her experience as worthwhile only in terms of bringing her closer to the completion of a project, so that she can move on to the next one, or move up the ranks, or move towards retirement. To live like this is arguably insane – but it's an insanity that gets inculcated in us early in life, as the self-styled 'spiritual entertainer' and New Age philosopher Alan Watts explained with characteristic vigour:

> Take education. What a hoax. As a child, you are sent to nursery school. In nursery school, they say you are getting ready to go on to kindergarten. And then first grade is coming up and second grade and third grade . . . In high school, they tell you you're getting ready for college.

> And in college you're getting ready to go out into the business world . . . [People are] like donkeys running after carrots that are hanging in front of their faces from sticks attached to their own collars. They are never here. They never get there. They are never alive.

The Causal Catastrophe

It took becoming a father for me to grasp how completely I'd spent my whole adult life, up to that point, mired in this future-chasing mindset. Not that the epiphany was instantaneous. Indeed, what happened first, as my son's birth approached, was that I became more obsessed than usual with using time well. Presumably every new parent, arriving home from the hospital to face the reality of their incompetence in the matter of child-rearing, feels some desire to spend their time as wisely as possible – first to keep the squirming bundle alive, and then to do whatever they can to lay the foundations for a happy future. But at the time, I was still enough of a productivity geek that I compounded my problems by purchasing several how-to books aimed at the parents of newborns; I was determined to make the very best use of those first crucial months.

This genre of publishing, I soon realised, was sharply divided into two camps, each in a permanent state of indignation at the mere existence of the other. On the one side were the gurus I came to think of as the Baby Trainers, who urged us to get our infant onto a strict schedule as soon as possible – because the absence of such structure would leave him existentially insecure and also because making his days more predictable would mean he could be seamlessly integrated into the rhythms of the household. This would allow everyone to get some sleep, and my wife and me to swiftly return to work. On the other side were the Natural Parents, for whom all such schedules – and frankly, the very notion of mothers having jobs to return to – were further evidence that modernity had corrupted the purity of parenthood, which could be recovered only by emulating the earthy practices of indigenous tribes in the developing world and/or prehistoric humans, these two groups being, to this camp of parenting experts, for all practical purposes, the same.

Later, I would learn that there's virtually no credible scientific evidence favouring either of these camps. (For example: the 'proof' that it's wrong to let your baby cry itself to sleep comes largely from research among infants abandoned in Romanian orphanages, which is hardly the same as leaving your child alone in their cosy Scandinavian bassinet for twenty minutes a day; meanwhile, there's one West African ethnic group, the Hausa-Fulani, who violate every Western parenting philosophy by deeming it taboo in some cases for mothers to make eye contact with their babies – and it seems those kids mostly turn out fine, too.) But what struck me most forcefully was how entirely preoccupied with the future both sets of experts were – indeed, how virtually all the parenting advice I encountered, in books and online, seemed utterly focused on doing whatever was required to produce the happiest or most successful or economically productive older children and adults later on.

This was obvious enough in the case of the Baby Trainers,

with their passion for inculcating good habits that might serve a baby well for life. But it was no less true of the Natural Parents. It would have been one thing if the Natural Parents' justifications for insisting on 'baby-wearing', or co-sleeping, or breastfeeding until age three had simply been that these were more satisfying ways for parents and babies to live. But their real motive, sometimes explicitly expressed, was that these were the best things to do to ensure a child's future psychological health. (Again: no real evidence.) And it struck me, rather more uncomfortably, that the reason I'd been seeking all this advice in the first place was because this was *my* stance on life, too: that for as long as I could remember, my days had been spent striving for future outcomes – exam results, jobs, better exercise habits: the list went on and on – in the service of some notional time when life would run smoothly at last. Now that my daily duties involved a baby, I'd simply expanded my instrumental approach to accommodate the new reality: I wanted to know that I was doing whatever was required to obtain optimal future results in the domain of child-rearing as well.

Except that this now began to seem to me like an astoundingly perverse way to approach spending time with a newborn, not to mention an unnecessarily exhausting thing to have to think about when life was already exhausting enough. Obviously, it mattered to keep half an eye on the future – there would be vaccinations to be administered, preschools to apply to, and so forth. But my son was here now, and he would be zero years old for only one year, and I came to realise that I didn't want to squander these days of his *actual* existence by focusing solely on how best to use them for the sake of his future one. He was sheer presence, participating unconditionally in the moment in which he found himself, and I wanted to join him in it. I wanted to watch his minuscule fist close around my finger, and his wobbly head turn in response to a noise, without obsessing over whether this showed he was meeting his 'developmental milestones' or not, or what I ought to be doing to ensure that he did. Worse still, it dawned on me that my fixation on using time well meant using my son himself, a whole other human being, as a tool for assuaging my own anxiety – treating him as nothing but a means to my hypothetical future sense of security and peace of mind.

The writer Adam Gopnik calls the trap into which I had fallen the 'causal catastrophe', which he defines as the belief 'that the proof of the rightness or wrongness of some way of bringing up children is the kind of adults it produces'. That idea sounds reasonable enough – how else would you judge rightness or wrongness? – until you realise that its effect is to sap childhood of any intrinsic value, by treating it as nothing but a training ground for adulthood. Maybe it really is a 'bad habit', as the Baby Trainers insist, for your one-year-old to grow accustomed to falling asleep on your chest. But it's also a delightful experience in the present moment, and that has to be weighed in the balance; it can't be the case that concerns for the future must always automatically take precedence. Likewise, the question of whether or not it's okay to let your nine-year-old spend hours each day playing violent video games doesn't turn solely on whether or not it'll turn him into a violent adult, but also on whether that's a good way for him to be using his life right now; perhaps a childhood immersed in digital blood and gore is just a lower-quality childhood, even if there aren't any future effects. In his play *The Coast of Utopia*, Tom Stoppard puts an intensified version of this sentiment into the mouth of the nineteenth-century Russian philosopher Alexander Herzen, as he struggles to come to terms with the death of his son, who has drowned in a shipwreck – and whose life, Herzen insists, was no less valuable for never coming to fruition in adult accomplishments. 'Because children grow up, we think a child's purpose is to grow up,' Herzen says. 'But a child's purpose is to be a child. Nature doesn't disdain what only lives for a day. It pours the whole of itself into each moment . . . Life's bounty is in its flow. Later is too late.'

The Last Time

And yet I hope it's clear by now that none of this applies only to people who happen to be the parents of small children. Certainly, it's true that a fast-developing newborn baby makes it especially hard to ignore the fact that life is a succession of transient experiences, valuable in themselves, which you'll miss if you're completely focused on the destination to which you hope they might be leading. But the author and podcast host Sam Harris makes the disturbing observation that the same applies to everything: our lives, thanks to their finitude, are inevitably full of activities that we're doing for the very last time. Just as there will be a final occasion on which I pick up my son – a thought that appals me, but one that's hard to deny, since I surely won't be doing it when he's thirty – there will be a last time that you visit your childhood home, or swim in the sea, or make love, or have a deep conversation with a certain close friend. Yet usually there'll be no way to know, in the moment itself, that you're doing it for the last time. Harris's point is that we should therefore try to treat every such experience with the reverence we'd show if it were the final instance of it. And indeed there's a sense in which every moment of life is a 'last time'. It arrives; you'll never get it again – and once it's passed, your remaining supply of moments will be one smaller than before. To treat all these moments solely as stepping stones to some future moment is to demonstrate a level of obliviousness to our real situation that would be jaw-dropping if it weren't for the fact that we all do it, all the time.

Admittedly, it's not entirely our own fault that we approach our finite time in such a perversely instrumental and future-focused way. Powerful external pressures push us in this direction, too, because we exist inside an economic system that is instrumentalist to its core. One way of understanding capitalism, in fact, is as a giant machine for instrumentalising everything it encounters – the earth's resources, your time and abilities (or 'human resources') – in the service of future profit. Seeing things this way helps explain the otherwise mysterious truth that rich people in capitalist economies are often surprisingly miserable. They're very good at instrumentalising their time, for the purpose of generating wealth for themselves; that's the definition of being successful in a capitalist world. But in focusing so hard on instrumentalising their time, they end up treating their lives in the present moment as nothing but a vehicle in which to travel towards a future state of happiness. And so their days are sapped of meaning, even as their bank balances increase.

This is also the kernel of truth in the cliché that people in less economically successful countries are better at enjoying life – which is another way of saying that they're less fixated on instrumentalising it for future profit, and are thus more able to participate in the pleasures of the present. Mexico, for example, has often outranked the United States in global indices of happiness. Hence the old parable about a vacationing New York businessman who gets talking to a Mexican fisherman, who tells him that he works only a few hours per day and spends most of his time drinking wine in the sun and playing music with his friends. Appalled at the fisherman's approach to time management, the businessman offers him an unsolicited piece of advice: if the fisherman worked harder, he explains, he could invest the profits in a bigger fleet of boats, pay others to do the fishing, make millions, then retire early. 'And what would I do then?' the fisherman asks. 'Ah, well, *then*,' the businessman replies, 'you could spend your days drinking wine in the sun and playing music with your friends.'

One vivid example of how the capitalist pressure towards instrumentalising your time saps meaning from life is the notorious case of corporate lawyers. The Catholic legal scholar Cathleen Kaveny has argued that the reason so many of them are so unhappy – despite being generally very well paid – is the convention of the 'billable hour', which obliges them to treat their time, and thus really themselves, as a commodity to be sold off in sixty-minute chunks to clients. An hour not sold is automatically an hour wasted. So when an outwardly successful, hard-driving lawyer fails to show up for a family dinner, or his child's school play, it's not necessarily because he's 'too busy', in the straightforward sense of having too much to do. It may also be because he's no longer able to conceive of an activity that can't be commodified as something worth doing at all. As Kaveny writes, 'Lawyers imbued with the ethos of the billable hour have difficulty grasping a non-commodified understanding of the meaning of time that would allow them to appreciate the true value of such participation.' When an activity can't be added to the running tally of billable hours, it begins to feel like an indulgence one can't afford. There may be more of this ethos in most of us – even the non-lawyers – than we'd care to admit.

And yet we'd be fooling ourselves to put all the blame on capitalism for the way in which modern life so often feels like a slog, to be 'got through' en route to some better time in the future. The truth is that we collaborate with this state of affairs. We *choose* to treat time in this self-defeatingly instrumental way, and we do so because it helps us maintain the feeling of being in omnipotent control of our lives. As long as you believe that the real meaning of life lies somewhere off in the future – that one day all your efforts will pay off in a golden era of happiness, free of all problems – you get to avoid facing the unpalatable reality that your life isn't leading towards some moment of truth that hasn't yet arrived. Our obsession with extracting the greatest future value out of our time blinds us to the reality that, in fact, the moment of truth is always now – that life is nothing but a succession of present moments, culminating in death, and that you'll probably never get to a point where you feel you have things in perfect working order. And that therefore you had better stop postponing the 'real meaning' of your existence into the future, and throw yourself into life now.

John Maynard Keynes saw the truth at the bottom of all this, which is that our fixation on what he called 'purposiveness' – on using time well for future purposes, or on 'personal productivity', he might have said, had he been writing today – is ultimately motivated by the desire not to die. 'The "purposive" man,' Keynes wrote, 'is always trying to secure a spurious and delusive immortality for his actions by pushing his interests in them forward into time. He does not love his cat, but his cat's kittens; nor in truth the kittens, but only the kittens' kittens, and so on forward forever to the end of cat-dom. For him, jam is not jam unless it is a case of jam tomorrow and never jam today. Thus by pushing his jam always forward into the future, he strives to secure for his act of boiling it an immortality.' Because he never has to 'cash out' the meaningfulness of his actions in the here and now, the purposive man gets to imagine himself an omnipotent god, whose influence over reality extends infinitely off into the future; he gets to feel as though he's truly the master of his time. But the price he pays is a steep one. He never gets to love an actual cat, in the present moment. Nor does he ever get to enjoy any actual jam. By trying too hard to make the most of his time, he misses his life.

Absent in the Present

Attempting to 'live in the moment', to find meaning in life *now*, brings its own challenges too, though. Have you ever actually tried it? Despite the insistence of modern mindfulness teachers that it's a speedy path to happiness – and despite a growing body of psychological research on the benefits of 'savouring', or making the deliberate effort to appreciate life's smaller pleasures – it turns out to be bewilderingly difficult to do. In his hippy classic *Zen and the Art of Motorcycle Maintenance*, Robert Pirsig describes arriving with his young son beside the blazing blue expanse of Crater Lake in Oregon, a collapsed prehistoric volcano that is America's deepest body of water. He's determined to get the most out of the experience, yet somehow he fails: '[We] see the Crater Lake with a feeling of "Well, there it is," just as the pictures show. I watch the other tourists, all of whom seem to have out-of-place looks too. I have no resentment at this, just a feeling that it's all unreal and that the quality of the lake is smothered by the fact that it's so pointed to.' The more you try to be here now, to point at what's happening in this moment and really see it, the more it seems like you *aren't* here now – or alternatively that you are, but that the experience has been drained of all its flavour.

I know how Pirsig must have felt. Several years ago, I visited Tuktoyaktuk, a small town in the extreme north of Canada's Northwest Territories. At the time, it was accessible only by air or sea or, in winter, by the route I took, which involved travelling in an off-road vehicle along the surface of a frozen river, past ships immobilised in ice for the season, and then driving upon the frozen Arctic Ocean itself. My journalistic assignment concerned the fight between Canada and Russia for oil resources beneath the North Pole – but naturally, having heard so much about them, I also wanted to see the northern lights. Several nights running, I forced myself outside into -30°C cold – a temperature at which the moisture inside your nose turns to ice the moment you inhale – to find only the darkness of thick cloud cover. It wasn't until my last night there, shortly after two o'clock in the morning, that the couple renting the neighbouring cabin at my bed-and-breakfast tapped excitedly on my door to tell me the time had come: the northern lights were on display. I threw some clothes over my full-body thermal underwear and stepped out under a cathedral sky, filled with moving curtains of green light, sweeping from horizon to horizon. I was determined to relish the exhibition, which the next morning locals would describe as a particularly impressive one. But the more I tried, the less I seemed able to do so. By the time I was getting ready to return to the warmth of my cabin, I was so far from being absorbed in the moment that a thought occurred to me, regarding the northern lights, which to this day I squirm to recall. *Oh*, I found myself thinking, *they look like one of those screen savers.*

The problem is that the effort to be present in the moment, though it seems like the exact opposite of the instrumentalist, future-focused mindset I've been criticising in this chapter, is in fact just a slightly different version of it. You're so fixated on trying to make the best use of your time – in this case not for some later outcome, but for an enriching experience of life right now – that it obscures the experience itself. It's like trying too hard to fall asleep, and therefore failing. You resolve to stay completely present while, say, washing the dishes – perhaps because you saw that quotation from the bestselling Buddhist teacher Thich Nhat Hanh about finding absorption in the most mundane of activities – only to discover that you can't, because you're too busy self-consciously wondering whether you're being present enough or not. The phrase 'be here now' calls to mind images of bearded stoners in bell-bottomed trousers, utterly relaxed about whatever's happening around them. Yet in fact the attempt to be here now feels not so much relaxing as rather strenuous – and it turns out that *trying* to have the most intense possible present-moment experience is a surefire way to fail. My favourite example of this effect is the 2015 study by researchers at Carnegie Mellon University in Pittsburgh, in which couples were instructed to have sex twice as frequently as usual for a two-month period. At the end of this time, the study concluded, they weren't any happier than they had been at the start. This finding was widely reported as demonstrating that a more active sex life isn't as enjoyable as you might have imagined. But what it really shows, I'd say, is that trying too hard to have a more active sex life is no fun at all.

A more fruitful approach to the challenge of living more fully in the moment starts from noticing that you are, in fact, always already living in the moment anyway, whether you like it or not. After all, your self-conscious thoughts about whether you're sufficiently focused on washing the dishes – or whether you're enjoying the extra sex you're having these days, since agreeing to participate in that psychology study – are thoughts arising in the present moment, too. And if you're inescapably already in the moment, there's surely something deeply dubious about trying to bring that state of affairs about. To *try* to live in the moment implies that you're somehow separate from 'the moment', and thus in a position to either succeed or fail at living in it. For all its chilled-out associations, the attempt to be here now is therefore still another instrumentalist attempt to use the present moment purely as a means to an end, in an effort to feel in control of your unfolding time. As usual, it doesn't work. The self-consciousness you experience when you seek too effortfully to be 'more in the moment' is the mental discomfort of attempting to lift yourself up by your own bootstraps – to modify your relationship to the present moment in time, when in fact that moment in time is all that you *are* to begin with.

As the American author Jay Jennifer Matthews puts it in her excellently titled short book *Radically Condensed Instructions for Being Just as You Are*, 'We cannot get anything out of life. There is no outside where we could take this thing to. There is no little pocket, situated outside of life, [to which we could] steal life's provisions and squirrel them away. The life of this moment has no outside.' Living more fully in the present may be simply a matter of finally realising that you never had any other option but to be here now.

9.

Rediscovering Rest

One boiling summer weekend a few years ago, I joined the impassioned members of a campaigning group called Take Back Your Time in an airless university lecture theatre in Seattle, where they'd assembled to further their long-standing mission of 'eliminating the epidemic of overwork'. The gathering in which I participated, their annual conference, was a sparsely attended affair – in part because, as the organisers conceded, it was August, and lots of people were on holiday, and America's most stridently pro-relaxation organisation could hardly complain about that. But it was also because Take Back Your Time promotes what counts, these days, as a highly subversive message. There's nothing unusual about its demands for more days off or shorter working hours; such proposals are increasingly common. But they are almost always justified on the grounds that a well-rested worker is a more productive worker – and it was precisely this rationale that the group had been set up to question. Why, its members wanted to know, should holidays by the sea, or meals with friends, or lazy mornings in bed need defending in terms of improved performance at work? 'You keep hearing people arguing that more time off might be good for the economy,' fumed John de Graaf, an ebullient seventy-ish film-maker and the driving force behind Take Back Your Time. 'But why should we have to justify *life* in terms of the *economy*? It makes no sense!' Later, I learned about the existence of a rival initiative, Project: Time Off, which, unlike Take Back Your Time, enjoyed generous corporate sponsorship, and better-attended conferences – and it was no surprise to learn that its mission was to promote 'the personal, business, social and economic benefits' of leisure. It was also backed by the US Travel Association, which has its own reasons for wanting people to take more holidays.

The Decline of Pleasure

De Graaf had put his finger on one of the sneakier problems with treating time solely as something to be used as well as possible, which is that we start to experience pressure to use our leisure time productively, too. Enjoying leisure for its own sake – which you might have assumed was the whole point of leisure – comes to feel as though it's somehow not quite enough. It begins to feel as though you're failing at life, in some indistinct way, if you're not treating your time off as an investment in your future. Sometimes this pressure takes the form of the explicit argument that you ought to think of your leisure hours as an opportunity to become a better worker ('Relax! You'll Be More Productive,' reads the headline on one hugely popular *New York Times* piece). But a more surreptitious form of the same attitude has also infected your friend who always seems to be training for a 10K, yet who's apparently incapable of just going for a run: she has convinced herself that running is a meaningful thing to do only insofar as it might lead towards a future accomplishment. And it infected me, too, during the years I spent attending meditation classes and retreats with the barely conscious goal that I might one day reach a condition of permanent calm. Even an undertaking as seemingly hedonistic as a year spent backpacking around the globe could fall victim to the same problem, if your purpose isn't to explore the world but – a subtle distinction, this – to add to your mental storeroom of experiences, in the hope that you'll feel, later on, that you'd used your life well.

The regrettable consequence of justifying leisure only in terms of its usefulness for other things is that it begins to feel vaguely like a chore – in other words, like work in the worst sense of that word. This was a pitfall the critic Walter Kerr noticed back in 1962, in his book *The Decline of Pleasure*: 'We are all of us compelled,' Kerr wrote, 'to read for profit, party for contacts . . . gamble for charity, go out in the evening for the greater glory of the municipality, and stay home for the weekend to rebuild the house.' Defenders of modern capitalism enjoy pointing out that despite how things might feel, we actually have more leisure time than we did in previous decades – an average of about five hours per day for men, and only slightly less for women. But perhaps one reason we don't experience life that way is that leisure no longer feels very leisurely. Instead, it too often feels like another item on the to-do list. And like many of our time troubles, research suggests that this problem grows worse the wealthier you get. Rich people are frequently busy working, but they also have more options for how to use any given hour of free time: like anyone else, they could read a novel or go for a walk; but they could equally be attending the opera, or planning a ski trip to Courchevel. So they're more prone to feeling that there are leisure activities they ought to be getting round to but aren't.

We probably can't hope to grasp how utterly alien this attitude towards leisure would have seemed to anyone living at any point before the Industrial Revolution. To the philosophers of the ancient world, leisure wasn't the means to some other end; on the contrary, it was the end to which everything else worth doing was a means. Aristotle argued that true leisure – by which he meant self-reflection and philosophical contemplation – was among the very highest of virtues because it was worth choosing for its own sake, whereas other virtues, like courage in war, or noble behaviour in government, were virtuous only because they led to something else. The Latin word for business, *negotium*, translates literally as 'not-leisure', reflecting the view that work was a deviation from the highest human calling. In this understanding of the situation, work might be an unavoidable necessity for certain people – above all, for the slaves whose toil made possible the leisure of the citizens of Athens and Rome – but it was fundamentally undignified, and certainly not the main point of being alive.

This same essential idea remained intact across centuries of subsequent historical upheaval: that leisure was life's centre of gravity, the default state to which work was a sometimes inevitable interruption. Even the onerous lives of medieval English peasants were suffused with leisure: they unfolded according to a calendar that was dominated by religious holidays and saints' days, along with multi-day village festivals, known as 'ales', to mark momentous occasions such as marriages and deaths. (Or less momentous ones, like the annual lambing, the season when ewes give birth – any excuse to get drunk.) Some historians claim that the average country-dweller in the sixteenth century would have worked for only about 150 days each year, and while those figures are disputed, nobody doubts that leisure lay near the heart of almost every life. Apart from anything else, while all that recreation might have been fun, it wasn't exactly optional. People faced strong social pressure not to work all the time: you observed religious holidays because the church required it; and in a close-knit village, it wouldn't have been easy to shirk the other festivities, either. Another result was that a sense of leisureliness seeped into the crevices of the days people did spend at work. 'The laboring man,' complained the Bishop of Durham, James Pilkington, sometime around 1570, 'will take his rest long in the morning; a good piece of the day is spent afore he comes at his work. Then he must have his breakfast, though he have not earned it, at his accustomed hour, else there is grudging and murmuring . . . At noon he must have his sleeping time, then his bever in the afternoon, which spendeth a great part of the day.'

But industrialisation, catalysed by the spread of the

clock-time mentality, swept all that away. Factories and mills required the coordinated labour of hundreds of people, paid by the hour, and the result was that leisure became sharply delineated from work. Implicitly, workers were offered a deal: you could do whatever you liked with your time off, so long as it didn't damage – and preferably enhanced – your usefulness on the job. (So there was a profit motive at play when the upper classes expressed horror at the lower classes' enthusiasm for drinking gin: coming to work with a hangover, because you'd spent your leisure time getting wasted, was a violation of the deal.) In one narrow sense, this new situation left working people freer than before, since their leisure was more truly their own than when church and community had dictated almost everything they did with it. But at the same time, a new hierarchy had been established. Work, now, demanded to be seen as the real point of existence; leisure was merely an opportunity for recovery and replenishment, for the purposes of further work. The problem was that for the average mill or factory worker, industrial work wasn't sufficiently meaningful to be the point of existence: you did it for the money, not for its intrinsic satisfactions. So now the whole of life – work and leisure time alike – was to be valued for the sake of something else, in the future, rather than for itself.

Ironically, the union leaders and labour reformers who campaigned for more time off, eventually securing the eight-hour workday and the two-day weekend, helped entrench this instrumental attitude towards leisure, according to which it could be justified only on the grounds of something other than pure enjoyment. They argued that workers would use any additional free time they might be given to improve themselves, through education and cultural pursuits – that they'd use it, in other words, for more than just relaxing. But there is something heartbreaking about the nineteenth-century Massachusetts textile workers who told one survey researcher what they actually longed to do with more free time: to 'look around to see what is going on'. They yearned for true leisure, not a different kind of productivity. They wanted what the maverick Marxist Paul Lafargue would later call, in the title of his best-known pamphlet, *The Right To Be Lazy*.

We have inherited from all this a deeply bizarre idea of what it means to spend your time off 'well' – and, conversely, what counts as wasting it. In this view of time, anything that doesn't create some form of value for the future is, by definition, mere idleness. Rest is permissible, but only for the purposes of recuperation for work, or perhaps for some other form of self-improvement. It becomes difficult to enjoy a moment of rest for itself alone, without regard for any potential future benefits, because rest that has no instrumental value feels wasteful.

The truth, then, is that spending at least some of your leisure time 'wastefully', focused solely on the pleasure of the experience, is the only way *not* to waste it – to be truly at leisure, rather than covertly engaged in future-focused self-improvement. In order to most fully inhabit the only life you ever get, you have to *refrain* from using every spare hour for personal growth. From this perspective, idleness isn't merely forgivable; it's practically an obligation. 'If the satisfaction of an old man drinking a glass of wine counts for nothing,' wrote Simone de Beauvoir, 'then production and wealth are only hollow myths; they have meaning only if they are capable of being retrieved in individual and living joy.'

Pathological Productivity

And yet here we'll need to confront a rarely acknowledged truth about rest, which is that we're not merely the victims of an economic system that denies us any opportunity for it. Increasingly, we're also the kind of people who don't actually want to rest – who find it seriously unpleasant to pause in our efforts to get things done, and who get antsy when we feel as though we're not being sufficiently productive. An extreme example is the case of the novelist Danielle Steel, who in a 2019 interview with *Glamour* magazine revealed the secret of how she'd managed to write 179 books by the time she turned seventy-two, releasing them at the rate of almost seven per year: by working almost literally all the time, in twenty-hour days, with a handful of twenty-four-hour writing periods per month, a single week's holiday each year, and practically no sleep. ('I don't get to bed until I'm so tired I could sleep on the floor,' she was quoted as saying. 'If I have four hours, it's a really good night for me.') Steel drew widespread praise for her 'badass' work habits. But it's surely not unreasonable to perceive, in this sort of daily routine, the evidence of a serious problem – of a deep-rooted inability to refrain from using time productively. In fact, Steel herself seems to concede that she uses productivity as a way to avoid confronting difficult emotions. Her personal ordeals have included the loss of an adult son to a drug overdose and no fewer than five divorces – and work, she told the magazine, is 'where I take refuge. Even when bad things have happened in my personal life, it's a constant. It's something solid I can escape into.'

If it seems uncharitable to accuse Steel of being pathologically unable to relax, I ought to clarify that the malady is widespread. I've suffered from it as acutely as anyone; and unlike Steel, I can't claim to have brought joy to millions of readers of romantic fiction as a happy side effect. Social psychologists call this inability to rest 'idleness aversion', which makes it sound like just another minor behavioural foible; but in his famous theory of the 'Protestant work ethic', the German sociologist Max Weber argued that it was one of the core ingredients of the modern soul. It first emerged, according to Weber's account, among Calvinist Christians in northern Europe, who believed in the doctrine of predestination – that every human, since before they were born, had been preselected to be a member of the elect, and therefore entitled to spend eternity in heaven with God after death, or else as one of the damned, and thus guaranteed to spend it in hell. Early capitalism got much of its energy, Weber argued, from Calvinist merchants and tradesmen who felt that relentless hard work was one of the best ways to prove – to others, but also to themselves – that they belonged to the former category rather than the latter. Their commitment to frugal living supplied the other half of Weber's theory of capitalism: when people spend their days generating vast amounts of wealth through hard work but also feel obliged not to fritter it away on luxuries, the inevitable result is large accumulations of capital.

It must have been a uniquely anguished way to live. There was no chance that all that hard work could improve the probability that one would be saved: after all, the whole point of the doctrine of predestination was that nothing could influence one's fate. On the other hand, wouldn't someone who already was saved naturally demonstrate a tendency towards virtuous striving and thriftiness? On this fraught understanding, idleness became an especially anxiety-inducing experience, to be avoided at all costs – not merely a vice that might lead to damnation if you overindulged in it, as many Christians had long maintained, but one that might be evidence of the horrifying truth that you already *were* damned.

We flatter ourselves that we've outgrown such superstitions today. And yet there remains, in our discomfort with anything that feels too much like wasting time, a yearning for something not all that dissimilar from eternal salvation. As long as you're filling every hour of the day with some form of striving, you get to carry on believing that all this striving is leading you somewhere – to an imagined future state of perfection, a heavenly realm in which everything runs smoothly, your limited time causes you no pain, and you're free of the guilty sense that there's more you need to be doing in order to justify your existence. Perhaps we shouldn't be too surprised when the activities with which we fill our leisure hours increasingly come to resemble not merely work but sometimes, as in the case of a SoulCycle class or a CrossFit workout, actual physical punishment – the self-flagellation of guilty sinners anxious to expunge the stain of laziness before it's too late.

To rest for the sake of rest – to enjoy a lazy hour for its own sake – entails first accepting the fact that this is it: that your days *aren't* progressing towards a future state of perfectly invulnerable happiness, and that to approach them with such an assumption is systematically to drain our four thousand weeks of their value. 'We are the sum of all the moments of our lives,' writes Thomas Wolfe, 'all that is ours is in them: we cannot escape it or conceal it.' If we're going to show up for, and thus find some enjoyment in, our brief time on the planet, we had better show up for it now.

Rules for Rest

Given all the blame I've been heaping on religion here for the modern Westerner's inability to relax, it might seem perverse to suggest that we should look to religion for the antidote as well. But it was members of religious communities who first understood a crucial fact about rest, which is that it isn't simply what occurs by default whenever you take a break from work. You need ways to make it likely that rest will actually happen.

Friends of mine live in an apartment building on the historically Jewish Lower East Side of New York that's equipped with a 'Shabbat elevator': step inside it between Friday evening and Saturday night, and you'll find yourself stopping at every floor, even if nobody wants to get on or off there, because it's been programmed to spare Jewish residents and visitors from having to violate the rule against operating electrical switches on the Sabbath. (In fact, the actual prohibition, laid down in ancient Jewish law, is against lighting fires, but modern authorities interpret that to include completing electrical circuits. The other thirty-eight categories of banned activities have been interpreted as outlawing everything from inflating water wings at the swimming pool to tearing sheets of toilet paper from a roll.) Such rules strike many of the rest of us as absurd. But if they are, it's an absurdity well tailored to an equally absurd reality about humans, which is that we need this sort of pressure in order to get ourselves to rest. As the writer Judith Shulevitz explains:

> Most people mistakenly believe that all you have to do to stop working is not work. The inventors of the Sabbath understood that it was a much more complicated undertaking. You cannot downshift casually and easily, the way you might slip into bed at the end of a long day. As the Cat in the Hat says, 'It is fun to have fun but you have to know how.' This is why the Puritan and Jewish Sabbaths were so exactingly intentional, requiring extensive advance preparation – at the very least a scrubbed house, a full larder and a bath. The rules did not exist to torture the faithful. They were meant to communicate the insight that interrupting the ceaseless round of striving requires a surprisingly strenuous act of will, one that has to be bolstered by habit as well as social sanction.

The idea of a communal weekly day off seems thoroughly old-fashioned today, persisting mainly in the memories of those older than about forty – who can still remember when most shops were open only six days per week – and in certain strange, vestigial laws, like the one in the city where I live prohibiting the purchase of alcohol before noon on Sundays.

As a result, we're in danger of forgetting what a radical notion the Sabbath always was – radical not least because, as the former slaves who inaugurated it were at pains to point out, it applied to everyone without exception. (Shulevitz notes that in the Torah verses setting out the rules for the Jewish Sabbath, the fact that even slaves must be allowed to rest gets mentioned twice, as if it were an alien idea, which the text's author knew would need to be forcefully driven home.) And since the dawn of capitalism, it's been radical in a second way: while capitalism gets its energy from the permanent anxiety of striving for more, the Sabbath embodies the thought that whatever work you've completed by the time that Friday (or Saturday) night rolls around might be *enough* – that there might be no sense, for now, in trying to get any more done. In his book *Sabbath as Resistance*, the Christian theologian Walter Brueggemann describes the Sabbath as an invitation to spend one day per week 'in the awareness and practice of the claim that we are situated on the receiving end of the gifts of God'. One need not be a religious believer to feel some of the deep relief in that idea of being 'on the receiving end' – in the possibility that today, at least, there might be nothing more you need to *do* in order to justify your existence.

All the same, it's surely never been harder to make the requisite psychological shift than it is today – to pause in your work for long enough to enter the coherent, harmonious, somehow thicker experience of time that comes with being 'on the receiving end' of life, the feeling of stepping off the clock into 'deep time', rather than ceaselessly struggling to master it. Societal pressures used to make it relatively easy to take time off: you couldn't go shopping when the shops weren't open, even if you wanted to, or work when the office was locked. Besides, you'd be much less likely to skip church, or Sunday lunch with the extended family, if you knew your absence would raise eyebrows. Now, though, the pressures all push us in the other direction: the shops are open all day, every day (and all night, online). And thanks to digital technology, it's all too easy to keep on working at home.

Personal or household rules, such as the increasingly popular idea of a self-imposed 'digital sabbath', can fill the vacuum to some extent. But they lack the social reinforcement that comes when everyone else is following the rule too, so they're inevitably harder to abide by – and because they're reliant on willpower, they're prone to all the hazards involved in trying to force yourself to be more 'present in the moment', as explored in the previous chapter. The other important thing we can do as individuals, in order to enter the experience of genuine rest, is simply to stop expecting it to feel good, at least in the first instance. 'Nothing is more alien to the present age than idleness,' writes the philosopher John Gray. He adds: 'How can there be play in a time when nothing has meaning unless it leads to something else?' In such an era, it's virtually guaranteed that truly stopping to rest – as opposed to training for a 10K, or heading off on a meditation retreat with the goal of attaining spiritual enlightenment – is initially going to provoke some serious feelings of discomfort, rather than of delight. That discomfort isn't a sign that you shouldn't be doing it, though. It's a sign that you definitely should.

Hiking as an End in Itself

It's just after half past seven on a rainy morning in midsummer when I park my car beside the road, zip up my waterproof jacket, and set off by foot into the high moors of the northern Yorkshire Dales. There's a splendour to this terrain that's most powerful when you're alone, and in no danger of being distracted from the barren drama of it all by pleasant conversation. So I'm happy to be solo as I head uphill, past a waterfall with a satisfyingly Satanic name – Hell Gill Force – and into open country, where the crunch of my walking boots sends startled grouse airborne from their hiding places in the heather. A mile or so further on, far from any road, I stumble on a tiny disused stone church with an unlocked door. The silence inside feels settled, as if it hasn't been disturbed in years, though in fact there were probably hikers here as recently as yesterday evening. Twenty minutes later, and I'm on the moor top, facing into the wind, savouring the bleakness I've always loved. I know there are people who'd prefer to be relaxing on a Caribbean beach, instead of getting drenched while trudging through gorse bushes under a glowering sky; but I'm not going to pretend I understand them.

Of course, this is just a country walk, perhaps the most mundane of leisure activities – and yet, as a way of spending one's time, it does have one or two features worth noting. For one thing, unlike almost everything else I do with my life, it's not relevant to ask whether I'm any good at it: all I'm doing is walking, a skill at which I haven't appreciably improved since around the age of four. Moreover, a country walk doesn't have a purpose, in the sense of an outcome you're trying to achieve or somewhere you're trying to get. (Even a walk to the supermarket has a goal – getting to the supermarket – whereas on a hike, you either follow a loop or reach a given point before turning back, so the most efficient way to reach the endpoint would be never to leave in the first place.) There are positive side effects, like becoming more physically fit, but that's not generally why people go on hikes. Taking a walk in the countryside, like listening to a favourite song or meeting friends for an evening of conversation, is thus a good example of what the philosopher Kieran Setiya calls an 'atelic activity', meaning that its value isn't derived from its telos, or ultimate aim. You shouldn't be aiming to get a walk 'done'; nor are you likely to reach a point in life when you've accomplished all the walking you were aiming to do. 'You can stop doing these things, and you eventually will, but you cannot *complete* them,' Setiya explains. They have 'no outcome whose achievement exhausts them and therefore brings them to an end'. And so the only reason to do them is for themselves alone: 'There is no more to going for a walk than what you are doing right now.'

As Setiya recalls in his book *Midlife*, he was heading towards the age of forty when he first began to feel a creeping sense of emptiness, which he would later come to understand as the result of living a project-driven life, crammed not with atelic activities but telic ones, the primary purpose of which was to have them done, and to have achieved certain outcomes. He published papers in philosophy journals in order to speed his path to academic tenure; he sought tenure in order to achieve a solid professional reputation and financial security; he taught students in order to achieve those goals, and also in order to help them attain degrees and launch their own careers. In other words, he was suffering from the very problem we've been exploring: when your relationship with time is almost entirely instrumental, the present moment starts to lose its meaning. And it makes sense that this feeling might strike in the form of a midlife crisis, because midlife is when many of us first become consciously aware that mortality is approaching – and mortality makes it impossible to ignore the absurdity of living solely for the future. Where's the logic in constantly postponing fulfilment until some later point in time when soon enough you won't have any 'later' left?

The most unsparingly pessimistic of philosophers, Arthur Schopenhauer, seems to have seen the emptiness of this sort of life as an unavoidable result of how human desire functions. We spend our days pursuing various accomplishments that we desire to achieve; and yet for any given accomplishment – attaining a permanent post at your university, say – it's always the case either that you haven't achieved it yet (so you're dissatisfied, because you don't yet have what you desire) or that you've already attained it (so you're dissatisfied, because you no longer have it as something to strive towards). As Schopenhauer puts it in his masterwork, *The World as Will and Idea*, it's therefore inherently painful for humans to have 'objects of willing' – things you want to do, or to have, in life – because not yet having them is bad, but getting them is arguably even worse: 'If, on the other hand, [the human animal] lacks objects of willing, because it is at once deprived of them again by too easy a satisfaction, a fearful emptiness and boredom comes over it; in other words, its being and its existence become an intolerable burden for it. Hence it swings like a pendulum to and fro between pain and boredom.' But the notion of the atelic activity suggests there's an alternative that Schopenhauer might have overlooked, one that hints at a partial solution to the problem of an overly instrumentalised life. We might seek to incorporate into our daily lives more things we do for their own sake alone – to spend some of our time, that is, on activities in which the only thing we're trying to get from them is the doing itself.

Rod Stewart, Radical

There's a less fancy term that covers many of the activities Setiya refers to as atelic: they are hobbies. His reluctance to use that word is understandable, since it's come to signify something slightly pathetic; many of us tend to feel that the person who's deeply involved in their hobby of, say, painting miniature fantasy figurines, or tending to their collection of rare cacti, is guilty of not participating in real life as energetically as they otherwise might. Yet it's surely no coincidence that hobbies have acquired this embarrassing reputation in an era so committed to using time instrumentally. In an age of instrumentalisation, the hobbyist is a subversive: he insists that some things are worth doing for themselves alone, despite offering no pay-offs in terms of productivity or profit. The derision we heap upon the avid stamp collector or train spotter might really be a kind of defence mechanism, to spare us from confronting the possibility that they're truly happy in a way that the rest of us – pursuing our telic lives, ceaselessly in search of future fulfilment – are not. This also helps explain why it's far less embarrassing (indeed, positively fashionable) to have a 'side hustle', a hobby-like activity explicitly pursued with profit in mind.

And so in order to be a source of true fulfilment, a good hobby probably *should* feel a little embarrassing; that's a sign you're doing it for its own sake, rather than for some socially sanctioned outcome. My respect for the rock star Rod Stewart increased a few years back when I learned – from newspaper coverage of an interview he'd given to *Railway Modeller* magazine – that he'd spent the last two decades at work on a vast and intricate model railway layout of a 1940s American city, a fantasy amalgam of New York and Chicago complete with skyscrapers, vintage automobiles and grimy sidewalks, with the grime hand-painted by Sir Rod himself. (He brought the layout on tour with him, requesting an additional hotel room to accommodate it.) Compare Stewart's hobby with, say, the kitesurfing antics of the entrepreneur Richard Branson. No doubt Branson sincerely finds kitesurfing enjoyable. But it's difficult not to interpret his choice of recreational activity as a calculated effort to enhance his brand as a daredevil – whereas Stewart's model train hobby is so at odds with his image as the leather-trousered, gravel-voiced singer of 'Do Ya Think I'm Sexy?' that it's impossible to avoid the conclusion he must genuinely do it out of love.

There's a second sense in which hobbies pose a challenge to our reigning culture of productivity and performance: it's fine, and perhaps preferable, to be mediocre at them. Stewart confessed to *Railway Modeller* that he isn't actually all that good at building model train layouts. (He paid someone else to do the fiddly electrical wiring.) But that might be part of why he enjoys it so much: to pursue an activity in which you have no hope of becoming exceptional is to put aside, for a while, the anxious need to 'use time well', which in Stewart's case presumably involves the need to keep on pleasing audiences, selling out stadiums, showing the world he's still got it. My other favourite pastime besides hiking – banging out the songs of Elton John on my electric piano – is so uplifting and absorbing, at least in part precisely because there's zero danger of my chimpanzee-level musicianship ever being rewarded with money or critical acclaim. By contrast, writing is a far more stressful undertaking, one in which it's harder to remain completely absorbed, because I can't eradicate the hope that I might accomplish it brilliantly, meeting with high praise or great commercial success, or at least do it well enough to shore up my sense of self-worth.

The publisher and editor Karen Rinaldi feels about surfing the same way that I do about cheesy piano rock, only more so: she dedicates every spare moment she can to it, and even wiped out her savings on a plot of land in Costa Rica for better access to the sea. Yet she readily admits that she remains an appalling surfer to this day. (It took her five years of attempting to catch a wave before she first managed to do so.) But 'in the process of trying to attain a few moments of bliss', Rinaldi explains, 'I experience something else: patience and humility, definitely, but also *freedom*. Freedom to pursue the futile. And the freedom to suck without caring is revelatory.' Results aren't everything. Indeed, they'd better not be, because results always come later – and later is always too late.

10.

The Impatience Spiral

If you've spent much time in a city where the honking of car horns is out of control – New York, say, or Mumbai – you'll know the special irritation of that sound, which derives from the fact that it isn't merely a disruption of the peace and quiet, but overwhelmingly a pointless disruption, too: it reduces everyone else's quality of life without improving the honker's. In my corner of Brooklyn, the evening rush-hour honking begins around 4 p.m. and continues until around eight; and in that stretch of time, there can't be more than a handful of honks in the entire borough that serve a practical purpose, like alerting someone to danger or rousing a driver who's failed to notice the light has changed. The message of all the other honks is simply 'Hurry up!' And yet every driver is stuck in the same traffic, with the same desire to make progress, and the same inability to do so; no sane honker can seriously believe that his honk will make the critical difference and get things moving at last. The pointless honk is thus symptomatic of another important way in which we're unwilling to acknowledge our limitations when it comes to our time: it's a howl of rage at the fact that the honker can't prod the world around him into moving as fast as he'd like it to.

That we suffer when we adopt this sort of dictatorial attitude towards the rest of reality is one of the central insights of the ancient Chinese religion of Taoism. The Tao Te Ching is full of images of suppleness and yielding: the wise man (the reader is constantly being informed) is like a tree that bends instead of breaking in the wind, or water that flows around obstacles in its path. Things just are the way they are, such metaphors suggest, no matter how vigorously you might wish they weren't – and your only hope of exercising any real influence over the world is to work with that fact, instead of against it. Yet the phenomenon of pointless honking, and of impatience more generally, suggests that most of us are pretty bad Taoists. We tend to feel as though it's our right to have things move at the speed we desire, and the result is that we make ourselves miserable – not just because we spend so much time feeling frustrated, but because chivvying the world to move faster is frequently counterproductive anyway. For example, traffic research long ago established that impatient driving behaviour tends to slow you down. (The practice of inching towards the car in front while waiting at a red light, a classic habit of the restless motorist, is wholly self-defeating – because once things start moving again, you have to accelerate more slowly than you otherwise would, so as to avoid rear-ending the vehicle ahead.) And the same goes for many of our other efforts to force reality's pace. Working too hastily means you'll make more errors, which you'll then be obliged to go back to correct; hurrying a toddler to get dressed, in order to leave the house, is all but guaranteed to make the process last much longer.

Escape Velocity

Though it's a hard thing to establish scientifically, we're almost certainly much more impatient than we used to be. Our decreasing tolerance for delay is reflected in statistics on everything from road rage and the length of politicians' sound bites to the number of seconds the average web user is prepared to wait for a slow-loading page. (It has been calculated that if Amazon's front page loaded one second more slowly, the company would lose $1.6 billion in annual sales.) And yet at first glance, as I mentioned in the introduction, this seems exceedingly strange. Virtually every new technology, from the steam engine to mobile broadband, has permitted us to get things done more quickly than before. Shouldn't this therefore have *reduced* our impatience, by allowing us to live at something closer to the speed we'd prefer? Yet since the beginning of the modern era of acceleration, people have been responding not with satisfaction at all the time saved but with increasing agitation that they can't make things move faster still.

This is another mystery, though, that's illuminated when you understand it as a form of resistance to our built-in human limitations. The reason that technological progress exacerbates our feelings of impatience is that each new advance seems to bring us closer to the point of transcending our limits; it seems to promise that *this* time, finally, we might be able to make things go fast enough for us to feel completely in control of our unfolding time. And so every reminder that in fact we *can't* achieve such a level of control starts to feel more unpleasant as a result. Once you can heat your dinner in the microwave in sixty seconds, it begins to seem genuinely realistic that you might be able to do so instantaneously, in zero seconds – and thus all the more maddeningly frustrating that you still have to wait an entire minute instead. (You'll have noticed how frequently the office microwave still has seven or eight seconds left on the clock from the last person who used it, a precise record of the moment at which the impatience became too much for them to bear.) Nor will it make much difference, unfortunately, if you personally manage to muster the inner serenity to avoid this kind of reaction, because you'll still end up suffering from *societal* impatience – that is, from the wider culture's rising expectations about how quickly things ought to happen. Once most people believe that one ought to be able to answer forty emails in the space of an hour, your continued employment may become dependent on being able to do so, regardless of your feelings on the matter.

There may be no more vivid demonstration of this ratcheting sense of discomfort, of wanting to hasten the speed of reality, than what's happened to the experience of reading. Over the last decade or so, more and more people have begun to report an overpowering feeling, whenever they pick up a book, that gets labelled 'restlessness' or 'distraction' – but which is actually best understood as a form of impatience, a revulsion at the fact that the act of reading takes longer than they'd like. 'I've been finding it harder and harder to concentrate on words, sentences, paragraphs,' laments Hugh McGuire, the founder of the public domain audiobook service LibriVox and (at least until recently) a lifelong reader of literary fiction. 'Let alone chapters. Chapters often have page after page of paragraphs.' He describes what's shifted in the formerly delicious experience of sliding into bed with a book: 'A sentence. Two sentences. Maybe three. And then . . . I needed just a *little* something else. Something to tide me over. Something to scratch the itch at the back of my mind – just a quick look at email on my iPhone; to write, and erase, a response to a funny tweet from William Gibson; to find, and follow, a link to a good, really good, article in the *New Yorker* . . .'

People complain that they no longer have 'time to read', but the reality, as the novelist Tim Parks has pointed out, is rarely that they literally can't locate an empty half-hour in the course of the day. What they mean is that when they do find a morsel of time, and use it to try to read, they find they're too impatient to give themselves over to the task. 'It is not simply that one is interrupted,' writes Parks. 'It is that one is actually *inclined* to interruption.' It's not so much that we're too busy, or too distractible, but that we're unwilling to accept the truth that reading is the sort of activity that largely operates according to its own schedule. You can't hurry it very much before the experience begins to lose its meaning; it refuses to consent, you might say, to our desire to exert control over how our time unfolds. In other words, and in common with far more aspects of reality than we're comfortable acknowledging, reading something properly just takes the time it takes.

Must Stop, Can't Stop

In the late 1990s, a psychotherapist in California named Stephanie Brown began to notice certain striking new patterns among the clients who came to seek her help. Brown's consulting rooms are in Menlo Park, in the heart of Silicon Valley, and as the first dot-com boom gathered steam, she found herself meeting its early casualties: well-paid, high-status overachievers who were so accustomed to a life of constant motion and stimulation that remaining seated for a fifty-minute therapy session seemed to cause them almost physical pain. It didn't take Brown long to figure out that their pulsing sense of urgency was a form of self-medication – something they were doing as a way not to feel something else. 'As soon as I slow down,' she remembers one woman telling her, in response to the suggestion that she might consider taking things a little more gently, 'the feeling of anxiety wells up inside, and I look for something to take it away.' Reaching for the smartphone, diving back into the to-do list, pounding away on the elliptical machine at the gym – all these forms of high-speed living were serving as some kind of emotional avoidance. As the months passed, it dawned on Brown that she recognised this sort of avoidance intimately herself. Her own experiences of it belonged to a life she'd long since left behind. But even so, the connection was clear: 'These people were talking about *exactly the same thing*!' she told me, the thrill of that initial realisation still audible in her voice. The high achievers of Silicon Valley reminded Brown of herself in her days as an alcoholic.

To understand the significance of this point, it helps to know that Brown, like many former drinkers, holds in high esteem the twelve-step philosophy of Alcoholics Anonymous, which asserts that alcoholism is fundamentally a result of attempting to exert a level of control over your emotions that you can't ever attain. The future alcoholic first turns to drink in an effort to escape some painful aspect of experience: Brown started drinking seriously at the age of sixteen, she said, because it seemed like the only way to banish a sense of yawning emotional distance between herself and her parents, both lifelong addicts themselves. 'I knew something was terribly wrong with our family from an early age,' she recalled, 'but when my father first offered me a glass of wedding champagne? I remember I was thrilled. No reflection at all. It was as if I finally got to join the family.'

At first this strategy seems to work, because drinking does temporarily numb unpleasant emotions. In the longer run, though, it backfires disastrously. Despite all your efforts to escape your experience, the truth is that you're still where you are – stuck in your dysfunctional family or your abusive relationship, suffering from depression, or not confronting the aftermath of childhood trauma – and so the feelings soon return, requiring stronger drinks in order to numb them. Only now, the alcoholic has additional problems: as well as struggling to control her emotions through drink, she must also try to control her drinking, lest it cost her her relationship, her job, or even her life. She'll probably start experiencing more friction at work and at home, and feel shame about her situation – all of which are triggers for further difficult emotions that are most easily numbed by more drink. This is the vicious spiral that constitutes the psychological core of an addiction. You know you *must* stop, but you also *can't* stop, because the very thing that's hurting you – alcohol – has come to feel like the only means of controlling the negative emotions that, in fact, your drinking is helping to cause.

Perhaps it seems melodramatic to compare 'addiction to speed', as Brown calls our modern disease of accelerated living, to a condition as serious as alcoholism. Some people definitely get offended when she does so. But her point isn't that compulsive hurry is as physically destructive as an excess of alcohol. It's that the basic mechanism is the same. As the world gets faster and faster, we come to believe that our happiness, or our financial survival, depends on our being able to work and move and make things happen at superhuman speed. We grow anxious about not keeping up – so to quell the anxiety, to try to achieve the feeling that our lives are under control, we move faster. But this only generates an addictive spiral. We push ourselves harder to get rid of anxiety, but the result is actually more anxiety, because the faster we go, the clearer it becomes that we'll never succeed in getting ourselves or the rest of the world to move as fast as we feel is necessary. (Meanwhile, we suffer the other effects of moving too fast: poor work output, a worse diet, damaged relationships.) Yet the only thing that feels feasible, as a way of managing all this additional anxiety, is to move faster still. You know you must stop accelerating, yet it also feels as though you can't.

This way of life isn't wholly unpleasant: just as alcohol gives the alcoholic a buzz, there's an intoxicating thrill to living at warp speed. (As the science writer James Gleick points out, it's no coincidence that another meaning of the word 'rush' is 'a feeling of exhilaration'.) But as a way of achieving peace of mind, it's doomed to fail. And whereas if you find yourself sliding into alcoholism, compassionate friends may try to intervene, to help steer you in the direction of a healthier life, speed addiction tends to be socially celebrated. Your friends are more likely to praise you for being 'driven'.

The futility of this situation – in which the addict's efforts to regain control send him spiralling further out of control – is the basis of the paradoxical-sounding insight for which Alcoholics Anonymous has become famous: that you can't truly hope to beat alcohol until you give up all hope of beating alcohol. This necessary shift in outlook generally happens as a result of 'hitting rock bottom', which is AA-speak for when things get so bad that you're no longer able to fool yourself. At that point, it becomes impossible for the alcoholic to avoid surrendering to the unpalatable truth of his limitations – to see that he simply doesn't have the ability to use alcohol as a strategic tool to suppress his most difficult emotions. ('We admitted,' reads the first of the Twelve Steps, [that] 'we were powerless over alcohol – that our lives had become unmanageable.') Only then, having abandoned the destructive attempt to achieve the impossible, can he get to work on what actually *is* possible: facing reality – above all, the reality that, in his case, there's no level of moderate drinking that's compatible with living a functioning life – then working, slowly and soberly, to fashion a more productive and fulfilling existence.

Likewise, Brown argues, we speed addicts must crash to earth. We have to give up. You surrender to the reality that things just take the time they take, and that you can't quiet your anxieties by working faster, because it isn't within your power to force reality's pace as much as you feel you need to, and because the faster you go, the faster you'll feel you need to go. If you can let those fantasies crumble, Brown's clients discovered, something unexpected happens, analogous to the alcoholic giving up his unrealistic craving for control in exchange for the gritty, down-to-earth, reality-confronting experience of recovery. Psychotherapists call it a 'second-order change', meaning that it's not an incremental improvement but a change in perspective that reframes everything. When you finally face the truth that you can't dictate how fast things go, you stop trying to outrun your anxiety, and your anxiety is transformed. Digging in to a challenging work project that can't be hurried becomes not a trigger for stressful emotions but a bracing act of choice; giving a difficult novel the time it demands becomes a source of relish. 'You cultivate an appreciation for endurance, hanging in, and putting the next foot forward,' Brown explains. You give up 'demanding instant resolution, instant relief from discomfort and pain, and magical fixes'. You breathe a sigh of relief, and as you dive into life as it really is, in clear-eyed awareness of your limitations, you begin to acquire what has become the least fashionable but perhaps most consequential of superpowers: patience.

11.

Staying on the Bus

It's fair to say that patience has a terrible reputation. For one thing, the prospect of doing anything that you've been told will require patience simply seems unappetising. More specifically, though, it's disturbingly passive. It is the virtue that has traditionally been urged upon housewives, while their husbands led more exciting lives outside the home; or on racial minorities, told to wait just a few more decades for their full civil rights. The talented but self-effacing employee who 'waits patiently' for a promotion, we tend to feel, will be waiting a long time: she ought to be trumpeting her achievements instead. In all such cases, patience is a way of psychologically accommodating yourself to a lack of power, an attitude intended to help you to resign yourself to your lowly position, in theoretical hopes of better days to come. But as society accelerates, something shifts. In more and more contexts, patience becomes a *form* of power.

In a world geared for hurry, the capacity to resist the urge to hurry – to allow things to take the time they take – is a way to gain purchase on the world, to do the work that counts, and to derive satisfaction from the doing itself, instead of deferring all your fulfilment to the future.

I first learned this lesson from Jennifer Roberts, who teaches art history at Harvard University. When you take a class with Roberts, your initial assignment is always the same, and it's one that has been known to elicit yelps of horror from her students: choose a painting or sculpture in a local museum, then go and look at it for three hours straight. No checking email or social media; no quick runs to Starbucks. (She reluctantly concedes that toilet breaks are allowed.) When I told a friend I planned to visit Harvard to meet Roberts, and to undertake the painting-viewing exercise myself, he gave me a look that mixed admiration with fear for my sanity, as though I'd announced an intention to kayak the Amazon alone. And he wasn't entirely wrong to worry about my mental health. There were long moments, as I squirmed in my seat at the Harvard Art Museum during the assignment, when I'd willingly have done countless things I usually can't stand – shopping for clothes, assembling flat-pack furniture, stabbing myself in the thigh with drawing pins – simply because I could have done them in a rush, instead of having to be patient.

Such reactions come as no surprise to Roberts. She insists on the exercise lasting three hours precisely because she knows it's a painfully long time, especially for anyone accustomed to a life of speed. She wants people to experience first hand how strangely excruciating it is to be stuck in position, unable to force the pace, and why it's so worthwhile to push past those feelings to what lies beyond. The idea first arose, Roberts told me, because her students faced so many external pressures to move fast – from digital technology, but also from Harvard's ultra-competitive atmosphere – that she began to feel it was insufficient for a teacher like her merely to hand out assignments and wait for the results. She felt she would be failing in her duties if she didn't also attempt to influence the tempo at which her students worked, helping them slow down to the speed that art demands. 'They needed someone to give them permission to spend this kind of time on *anything*,' she said. 'Somebody had to give them a different set of rules and constraints than the ones that were dominating their lives.'

Certain art forms impose temporal restraints on their audience in a rather obvious way: when you watch, say, a live performance of *The Marriage of Figaro* or a screening of *Lawrence of Arabia*, you don't have much choice but to let the work in question take its time. But other kinds, including painting, benefit from external restraints – because it's all too easy to tell yourself that once you've taken a couple of seconds to *look* at a painting, you've thereby genuinely *seen* it. So to prevent her students from rushing the assignment, Roberts had to make 'not rushing' the assignment itself.

She undertook the exercise herself, too, with a painting called *Boy with a Squirrel*, by the American artist John Singleton Copley. (It shows a boy with a squirrel.) 'It took me nine minutes to notice that the shape of the boy's ear precisely echoes that of the ruff along the squirrel's belly,' Roberts later wrote, 'and that Copley was making some kind of connection between the animal and the human body . . . It took a good 45 minutes before I realized that the seemingly random folds and wrinkles in the background curtain were actually perfect copies of the boy's ear and eye.'

There is nothing passive or resigned about the kind of patience that arises from this effort to resist the urge to hurry. On the contrary, it's an active, almost muscular state of alert presence – and its benefits, as we'll see, extend far beyond art appreciation. But for the record, here is what happens when you spend three unbroken hours on a small fold-out seat at the Harvard Art Museum looking at *Cotton Merchants in New Orleans*, a painting by Edgar Degas, with your phone, laptop and other distractions stowed out of reach in the cloakroom: you spend the first forty minutes wondering what on earth you'd been thinking. You remember – how could you ever have forgotten? – that you've always hated art galleries, especially the way their shuffling crowds of visitors impart a sort of contagious lethargy to the air. You contemplate switching paintings, from a work that now strikes you as a self-evidently tedious choice (it shows three men, in a room, inspecting some bales of cotton) to a nearby alternative, which seems to show many tiny souls being tortured in hell. But then you're forced to admit to yourself that making a fresh start, by picking a new painting, would be to succumb to the very impatience you're here to learn to resist – an attempt to seize control over your experience in precisely the way you're seeking to avoid. And so you wait. Grumpiness gives way to fatigue, then restless irritation. Time slows and sags. You wonder if an hour has passed, but when you check your watch, you find it's been seventeen minutes.

And then, around the eighty-minute mark, but without your noticing precisely when or how it happens, there's a shift. You finally give up attempting to escape the discomfort of time passing so slowly, and the discomfort abates. And the Degas begins to reveal its secret details: subtle expressions of watchfulness and sadness on the faces of the three men – one of whom, you notice properly for the first time, is a black merchant in an otherwise white milieu – plus an unexplained shadow you hadn't previously seen, as if a fourth person were lurking out of view; and a curious optical illusion that renders one of the figures either conventionally solid or transparent, like a ghost, depending on how your eyes interpret the painting's other lines. Before long, you're experiencing the scene in all its sensory fullness: the humidity and claustrophobia of that room in New Orleans, the creak of the floorboards, the taste of dust in the air.

The second-order change has occurred: now that you've abandoned your futile efforts to dictate the speed at which the experience moves, the real experience can begin. And you start to understand what the philosopher Robert Grudin means when he describes the experience of patience as 'tangible, almost edible', as if it gives things a kind of chewiness – the word is inadequate, but it's the closest one there is – into which you can sink your teeth. Your reward for surrendering the fantasy of controlling the pace of reality is to achieve, at last, a real sense of purchase on that reality – of really *getting stuck in* to life.

Watching and Waiting

In his book *The Road Less Travelled*, the psychotherapist M. Scott Peck recounts a transformative experience of surrendering to the speed of reality – one that emphasises that patience isn't merely a more peaceful and present-orientated way to live but a concretely useful skill. Until the age of thirty-seven, Peck explains, he considered himself a 'mechanical idiot', almost entirely inept when it came to fixing household appliances, cars, bicycles and suchlike. Then one day he came upon a neighbour who was midway through fixing his lawnmower, and paid him a self-deprecating compliment: 'Boy, I sure admire you. I've never been able to fix those kinds of things!'

'That's because you don't take the time,' the neighbour replied – a comment that gnawed at Peck, troubling something in his soul, and that resurfaced a few weeks later when the handbrake on a car belonging to one of his therapy patients became stuck. Normally, he writes, he would have 'immediately yanked at a few wires without having the foggiest idea of what I was doing, and then, when nothing constructive resulted, would have thrown up my hands and proclaimed "It's beyond me!"' This time, though, Peck remembered his neighbour's admonition:

> I lay down on the floor beneath the front seat of [the] car. Then I took the time to make myself comfortable. Once I was comfortable, I then took the time to look at the

> situation . . . At first all I saw was a confusing jumble of wires and tubes and rods, whose meaning I did not know. But gradually, in no hurry, I was able to focus my sight on the brake apparatus and trace its course. And then it became clear to me that there was a little latch preventing the brake from being released. I slowly studied this latch until it became clear to me that if I were to push it upward with the tip of my finger, it would move easily and would release the brake. And so I did this. One single motion, one ounce of pressure from a fingertip, and the problem was solved. I was a master mechanic!

Peck's insight here – that if you're willing to endure the discomfort of not knowing, a solution will often present itself – would be helpful enough if it were merely a piece of advice for fixing lawnmowers and cars. But his larger point is that it applies almost everywhere in life: to creative work and relationship troubles, politics and parenting. We're made so uneasy by the experience of allowing reality to unfold at its own speed that when we're faced with a problem, it feels better to race towards a resolution – any resolution, really, so long as we can tell ourselves we're 'dealing with' the situation, thereby maintaining the feeling of being in control. So we snap at our partners, rather than hearing them out, because waiting and listening would make us feel – correctly – as though we weren't in control of the situation. Or we abandon difficult creative projects, or nascent romantic relationships, because there's less uncertainty in just calling things off than in waiting to see how they might develop. Peck recalls one patient, an accomplished financial analyst in her professional life, who took this same rushed approach to the challenge of disciplining her children: 'Either she made the very first change that came to her mind within a matter of seconds – making them eat more breakfast or sending them to bed earlier – regardless of whether such a change had anything to do with the problem, or else she came to her next therapy session . . . despairing: "It's beyond me. What shall I do?"'

Three Principles of Patience

In practical terms, three rules of thumb are especially useful for harnessing the power of patience as a creative force in daily life. The first is to *develop a taste for having problems.* Behind our urge to race through every obstacle or challenge, in an effort to get it 'dealt with', there's usually the unspoken fantasy that you might one day finally reach the state of having no problems whatsoever. As a result, most of us treat the problems we encounter as doubly problematic: first because of whatever specific problem we're facing; and second because we seem to believe, if only subconsciously, that we shouldn't have problems at all. Yet the state of having no problems is obviously never going to arrive. And more to the point, you wouldn't want it to, because a life devoid of all problems would contain nothing worth doing, and would therefore be meaningless. Because what is a 'problem', really? The most generic definition is simply that it's something that demands that you address yourself to it – and if life contained no such demands, there'd be no point in anything. Once you give up on the unattainable goal of eradicating all your problems, it becomes possible to develop an appreciation for the fact that life just *is* a process of engaging with problem after problem, giving each one the time it requires – that the presence of problems in your life, in other words, isn't an impediment to a meaningful existence but the very substance of one.

The second principle is to *embrace radical incrementalism*. The psychology professor Robert Boice spent his career studying the writing habits of his fellow academics, reaching the conclusion that the most productive and successful among them generally made writing a *smaller* part of their daily routine than the others, so that it was much more feasible to keep going with it day after day. They cultivated the patience to tolerate the fact that they probably wouldn't be producing very much on any individual day, with the result that they produced much more over the long term. They wrote in brief daily sessions – sometimes as short as ten minutes, and never longer than four hours – and they religiously took weekends off. The panicked PhD students in whom Boice tried to inculcate this regimen rarely had the forbearance to hear it. They had looming deadlines, they protested, and couldn't afford such self-indulgent work habits. They needed their dissertations finished, and fast! But for Boice, that reaction just proved his point. It was precisely the students' impatient desire to hasten their work beyond its appropriate pace, to race on to the point of completion, that was impeding their progress. They couldn't stand the discomfort that arose from being forced to acknowledge their limited control over the speed of the creative process – and so they sought to escape it, either by not getting down to work at all, or by rushing headlong into stressful all-day writing binges, which led to procrastination later on, because it made them learn to hate the whole endeavour.

One critical aspect of the radical incrementalist approach, which runs counter to much mainstream advice on productivity, is thus to be willing to stop when your daily time is up, even when you're bursting with energy and feel as though you could get much more done. If you've decided to work on a given project for fifty minutes, then once fifty minutes have elapsed, get up and walk away from it. Why? Because as Boice explained, the urge to push onward beyond that point 'includes a big component of impatience about not being finished, about not being productive enough, about never again finding such an ideal time' for work. Stopping helps strengthen the muscle of patience that will permit you to return to the project again and again, and thus to sustain your productivity over an entire career.

The final principle is that, more often than not, *originality lies on the far side of unoriginality.* The Finnish American photographer Arno Minkkinen dramatises this deep truth about the power of patience with a parable about Helsinki's main bus station. There are two dozen platforms there, he explains, with several different bus lines departing from each one – and for the first part of its journey, each bus leaving from any given platform takes the same route through the city as all the others, making identical stops. Think of each stop as representing one year of your career, Minkkinen advises photography students. You pick an artistic direction – perhaps you start working on platinum prints of nudes – and you begin to accumulate a portfolio of work. Three years (or bus stops) later, you proudly present it to the owner of a gallery. But you're dismayed to be told that your pictures aren't as original as you thought, because they look like knock-offs of the work of the photographer Irving Penn; Penn's bus, it turns out, had been on the same route as yours. Annoyed at yourself for having wasted three years following somebody else's path, you jump off that bus, hail a taxi, and return to where you started at the bus station. This time, you board a different bus, choosing a different genre of photography in which to specialise. But a few stops later, the same thing happens: you're informed that your new body of work seems derivative, too. Back you go to the bus station. But the pattern keeps on repeating: nothing you produce ever gets recognised as being truly your own.

What's the solution? 'It's simple,' Minkkinen says. 'Stay on the bus. *Stay on the fucking bus.*' A little further out on their journeys through the city, Helsinki's bus routes diverge, plunging off to unique destinations as they head through the suburbs and into the countryside beyond. *That's* where the distinctive work begins. But it begins at all only for those who can muster the patience to immerse themselves in the earlier stage – the trial-and-error phase of copying others, learning new skills and accumulating experience.

The implications of this insight aren't confined to creative work. In many areas of life, there's strong cultural pressure to strike out in a unique direction – to spurn the conventional options of getting married, or having kids, or remaining in your home town, or taking an office job, in favour of something apparently more exciting and original. Yet if you always pursue the unconventional in this way, you deny yourself the possibility of experiencing those other, richer forms of uniqueness that are reserved for those with the patience to travel the well-trodden path first. As in Jennifer Roberts's three-hour painting-viewing exercise, this begins with the willingness to stop and be where you are – to engage with that part of the journey, too, instead of always badgering reality to hurry up. To experience the profound mutual understanding of the long-married couple, you have to stay married to one person; to know what it's like to be deeply rooted in a particular community and place, you have to stop moving around. Those are the kinds of meaningful and singular accomplishments that just take the time they take.

12.

The Loneliness of the Digital Nomad

Patience isn't the only way in which it's possible to find a deeper sort of freedom in surrendering to temporal constraints, instead of always trying to dictate how things unfold. Another concerns the perpetually irritating phenomenon of other human beings – who, as I take it you've noticed, are always impinging on your time in countless frustrating ways. It's a principle common to virtually all productivity advice that, in an ideal world, the only person making decisions about your time would be you: you'd set your own hours, work wherever you chose, take holidays when you wished, and generally be beholden to nobody. But there's a case to be made that this degree of control comes at a cost that's ultimately not worth paying.

Whenever I'm feeling resentful about deadlines, or the toddler's unpredictable sleep patterns, or other incursions upon my temporal sovereignty, I try to remember the cautionary tale of Mario Salcedo, a Cuban American financial consultant who almost certainly holds the record for the number of nights spent aboard cruise ships. There's little question that Super Mario – as he's known to the staff of Royal Caribbean Cruises, the firm to which he's been loyal for most of his two decades, as a resident of the oceans, with the 2020 coronavirus pandemic the only major interruption – is in full control of his time. 'I don't have to take out the garbage, I don't have to clean, I don't have to do laundry – I've eliminated all those non-value-added activities, and just have all the time in the world to enjoy what I like to do,' he once told the film maker Lance Oppenheim, poolside on board the *Enchantment of the Seas*. But it'll come as no surprise, presumably, to learn that he doesn't seem all that happy. In Oppenheim's short film, *The Happiest Guy in the World*, Salcedo prowls the decks, cocktail in hand, staring out to sea, eliciting tight-lipped smiles and reluctant pecks on the cheek from the people he refers to as his 'friends' – the employees of Royal Caribbean Cruises – and complaining that he can't get Fox News on the television in his cabin. 'I'm probably the happiest guy in the world!' he informs random groups of other passengers, rather too insistently; and they smile and nod, and politely pretend that they envy him.

Of course, it's not my place to assert that Salcedo isn't as happy as he claims. Perhaps he is. But I do know that *I* wouldn't be, were I living his life. The problem, I think, is that his lifestyle is predicated on a misunderstanding about the value of time. To borrow from the language of economics, Salcedo sees time as a regular kind of 'good' – a resource that's more valuable to you the more of it you command. (Money is the classic example: it's better to control more of it than less.) Yet the truth is that time is also a 'network good', one that derives its value from how many *other* people have access to it, too, and how well their portion is coordinated with yours. Telephone networks are the obvious example here: telephones are valuable to the extent that others also have them. (The more people who own phones, the more beneficial it is for you to own one; and unlike money, there's little point in accumulating as many phones as possible for your personal use.) Social media platforms follow the same logic. What matters isn't how many Facebook profiles you have, but that others have them, too, and that they're linked to yours.

As with money, it's good to have plenty of time, all else being equal. But having all the time in the world isn't much use if you're forced to experience it all on your own. To do countless important things with time – to socialise, go on dates, bring up children, launch businesses, build political movements, make technological advances – it has to be synchronised with other people's. In fact, having large amounts of time but no opportunity to use it collaboratively isn't just useless but actively unpleasant – which is why, for premodern people, the worst of all punishments was to be physically ostracised, abandoned in some remote location where you couldn't fall in with the rhythms of the tribe. And yet in achieving so much dominion over his time, Super Mario seems to have imposed a slightly milder version of the same fate on himself.

In and Out of Sync

The truly troubling thought, though, is that those of us who'd never dream of choosing a lifestyle like Salcedo's might nonetheless be guilty of the same basic mistake – of treating our time as something to hoard, when it's better approached as something to share, even if that means surrendering some of your power to decide exactly what you do with it and when. The quest for more individual control over my time, I have to admit, was a major motivation behind my decision to leave my job at a newspaper to become a work-from-home writer. And it's the implicit rationale behind many workplace policies we tend to think of as unquestionably good, such as parent-friendly flexitime and arrangements that give employees the option of working remotely, which seem certain to become much more common after the experience of lockdown during the pandemic. 'A person with a flexible schedule and average resources will be happier than a rich person who has everything except a flexible schedule,' advises the cartoonist turned self-help guru Scott Adams, summarising the ethos of individual time sovereignty. And so, he goes on, 'step one in your search for happiness is to continually work toward having control of your schedule'. The most extreme expression of this outlook is the modern lifestyle choice of becoming a 'digital nomad' – someone who liberates herself from the rat race in order to travel the globe with her laptop, operating her internet business from a Guatemalan beach or Thai mountaintop, as her fancy dictates.

But 'digital nomad' is a misnomer – and an instructive one. Traditional nomads aren't solitary wanderers who just happen to lack laptops; they're intensely group-focused people who, if anything, have *less* personal freedom than members of settled tribes, since their survival depends on their working together successfully. And in their more candid moments, digital nomads will admit that the chief problem with their lifestyle is the acute loneliness. 'Last year, I visited 17 countries; this year, I will visit 10,' the American author Mark Manson wrote, back when he was still a nomad himself. 'Last year, I saw the Taj Mahal, the Great Wall of China and Machu Picchu in the span of three months . . . But I did all this alone.' A fellow wanderer, Manson learned, 'burst into tears in a small suburb in Japan watching families ride their bikes together in a park', as it dawned on him that his supposed freedom – his theoretical ability to do whatever he wanted, whenever he chose – had put such ordinary pleasures beyond reach.

The point, to be clear, isn't that freelancing or long-term travel – let alone family-friendly workplace policies – are intrinsically bad things. It's that they come with an unavoidable flip side: every gain in personal temporal freedom entails a corresponding loss in how easy it is to coordinate your time with other people's. The digital nomad's lifestyle lacks the shared rhythms required for deep relationships to take root. For the rest of us, likewise, more freedom to choose when and where you work makes it harder to forge connections through your job, as well as less likely you'll be free to socialise when your friends are.

In 2013, a researcher from Uppsala in Sweden named

Terry Hartig, along with several colleagues, elegantly proved the connection between synchronisation and life satisfaction when he had the ingenious notion of comparing Swedes' holiday patterns against statistics on the rate at which pharmacists dispensed antidepressants. One of his two central findings was unremarkable: when Swedes take time off work, they're happier (as measured by their being less likely, on average, to need antidepressants). But the other was revelatory: antidepressant use fell by a greater degree, Hartig demonstrated, in proportion to how much of the population of Sweden was on holiday at any given time. Or to put things slightly differently, the more Swedes who were off work simultaneously, the happier people got. They derived psychological benefits not merely from holiday time, but from having the same holiday time as other people. When many were on holiday at once, it was as if an intangible, supernatural cloud of relaxation had settled over the nation as a whole.

Except that when you think about it, this makes perfect, non-supernatural sense. It's much easier to nurture relationships with family and friends when they're off work, too. Meanwhile, if you can be sure the whole office is deserted while you're trying to relax, you're spared the anxiety of thinking about all the undone tasks that might be accumulating, the emails filling up your inbox, or the scheming colleagues attempting to steal your job. Nonetheless, there was something a little spooky about how widely the beneficial effects of synchronised holidays rippled through the country. Hartig showed that even retired people, despite not having jobs from which to take a break, were happier when more of the Swedish workforce was on holiday. That finding echoed other research, which has demonstrated that people in long-term unemployment get a happiness boost when the weekend arrives, just like employed people relaxing after a busy workweek, though they don't have a workweek in the first place. The reason is that part of what makes weekends fun is getting to spend time with others who are also off work – plus, for the unemployed, the weekend offers respite from feelings of shame that they ought to be working when they aren't.

Hartig did not flinch from the controversial implication of his results. They suggest, he observed, that what people need isn't greater individual control over their schedules but rather what he calls 'the social regulation of time': greater outside pressure to use their time in particular ways. That means more willingness to fall in with the rhythms of community; more traditions like the Sabbath of decades past, or the French phenomenon of the *grandes vacances*, where almost everything grinds to a halt for several weeks each summer. Perhaps it even means more laws to regulate when people can and can't work, like limits on Sunday opening hours or recent European legislation banning certain employers from sending work emails out of hours.

On a work trip to Sweden a few years ago, I experienced a micro-level version of the same idea in the form of the *fika*, the daily moment when everyone in a given workplace gets up from their desks to gather for coffee and cake. The event resembles a well-attended coffee break, except that Swedes are liable to become mildly offended – which is the equivalent of a non-Swede becoming severely offended – if you suggest that's all it is. Because something intangible but important happens at the *fika*. The usual divisions get set aside; people mingle without regard for age, or class, or status within the office, discussing both work-related and non-work matters: for half an hour or so, communication and conviviality take precedence over hierarchy and bureaucracy. One senior manager told me it was by far the most effective way to learn what was really going on in his company. Yet it works only because those involved are willing to surrender some of their individual sovereignty over their time. You can choose to pause for coffee at some other time instead, if you insist. But eyebrows may be raised.

The other way to grasp how greatly we benefit from surrendering to communal time – whether we realise it or not – is to watch what happens when people are forcibly prevented from doing so. The historian Clive Foss has described the nightmare that transpired when the leadership of the Soviet Union, gripped by the desire to transform the nation into one blazingly efficient machine, set out to re-engineer time itself. The Soviets had long been inspired by the work of the efficiency expert Frederick Winslow Taylor, whose philosophy of 'scientific management' had aimed to squeeze the maximum possible output from American factory workers. But now Josef Stalin's chief economist, Yuri Larin, concocted what seems in hindsight like a ludicrously ambitious plan to keep Soviet factories running every day of the year, with no breaks. Henceforth, he announced in August 1929, a week would be not seven but five days long: four days of work, followed by one day's rest. Crucially, though, the idea was that not all workers would follow the same calendar.

Instead, they'd be divided into five groups, identified by a colour – yellow, green, orange, purple, red – each of which would then be assigned a different four-day workweek and one-day weekend, so that operations would never have to cease, even for a day. Meanwhile, Soviet authorities argued, there would be numerous benefits for the proletariat, too: more frequent days off, plus less overcrowding of cultural institutions and supermarkets, thanks to the steadier flow of customers.

But the main effect for ordinary citizens of the USSR, as the writer Judith Shulevitz has explained, was to destroy the possibility of social life. It was a simple question of scheduling. Two friends assigned to two different calendar groups would never be free to socialise on the same day. Husbands and wives were supposed to be assigned to the same group, but they often weren't, placing intense stress on families; and for obvious reasons, Sunday religious gatherings were disrupted, too – neither of which posed a problem from Moscow's point of view, since it was part of communism's mission to undermine the rival power centres of family and church. (E. G. Richards, a historian who chronicled the experiment, noted that 'Lenin's widow, in good Marxist fashion, regarded Sunday family reunions as a good enough reason to abolish that day'.) As one worker rather daringly complained to the official newspaper *Pravda*: 'What are we to do at home if the wife is in the factory, the children in school, and no one can come to see us? What is left but to go to the public tea room? What kind of life is that, when holidays come in shifts, and not for all workers together? That's no holiday, if you have to celebrate it by yourself.' The restructured workweek persisted in some form until 1940, when it was abandoned because of problems it caused with the maintenance of machinery. But not before the Soviet government had inadvertently demonstrated how much of the value of time comes not from the sheer quantity you have, but from whether you're in sync with the people you care about most.

Keeping Together in Time

There is an even more visceral sense, as well, in which time just feels *realer* – more intense, more vivid, more filled with meaning – when you're synchronised well with others. In 1941, a young American named William McNeill was drafted into the United States Army and sent for basic training at a camp on a dusty expanse of Texas scrubland. Nominally, his task was to learn how to fire anti-aircraft guns, but since the camp had only one such gun among thousands of trainees, and since even that gun wasn't fully functional, the officers in charge filled the long stretches of empty time with traditional military marching drills instead. On the face of it, as even a novice like McNeill understood, such exercises were completely pointless: by the time of the Second World War, troops were being transported across large distances in trucks and trains, not on foot; and in the era of the machine gun, to engage in formal marching in the heat of battle itself was essentially asking the enemy to slaughter you. And so McNeill was unprepared for the way in which the experience of marching with his fellow soldiers overwhelmed him:

> Marching aimlessly about on the drill field, swaggering in conformity with prescribed military postures, conscious only of keeping in step so as to make the next move correctly and in time somehow felt good. Words are inadequate to describe the emotion aroused by the prolonged movement in unison that drilling involved. A sense of pervasive well-being is what I recall; more specifically, a strange sense of personal enlargement; a sort of swelling out, becoming bigger than life, thanks to participation in collective ritual . . . Moving briskly and keeping in time was enough to make us feel good about ourselves, satisfied to be moving together, and vaguely pleased with the world at large.

The experience stuck with McNeill, and after the war, when he became a professional historian, he returned to the idea in a monograph called *Keeping Together in Time*. In it, he argues that synchronised movement, along with synchronised singing, has been a vastly underappreciated force in world history, fostering cohesion among groups as diverse as the builders of the pyramids, the armies of the Ottoman Empire, and the Japanese office workers who rise from their desks to perform group calisthenics at the start of each workday. Roman generals were among the first to discover that soldiers marching in synchrony could be made to travel for far longer distances before they succumbed to fatigue. And some evolutionary biologists speculate that music itself – a phenomenon that has proved difficult to account for in terms of Darwinian natural selection, except as a pleasurable by-product of more important mechanisms – might have emerged as a way of coordinating large groups of tribal warriors, who could move in unison by following rhythms and melodies, where other forms of communication would have proved too cumbersome for the job.

In daily life, as well, we fall into synchrony all the time, usually without realising it: at the theatre, applause gradually organises itself into a rhythm; and if you walk down the street alongside a friend, or even a stranger, you'll soon find your paces starting to match. This subliminal urge towards coordinated action is so powerful that even sworn rivals can't resist it. It would be difficult to imagine two men more committed to defeating each other – on a conscious level, at least – than the sprinters Usain Bolt and Tyson Gay, competing for the men's hundred-metre title at the World Athletics Championships in 2009. But a study based on a frame-by-frame analysis of the race shows that despite presumably being consumed by the desire to win, Bolt couldn't help falling into line with Gay's steps. And Bolt almost certainly benefited as a result: other research has indicated that conforming to an external rhythm renders one's gait imperceptibly more efficient. So it's likely that Gay, in spite of himself, helped his opponent to reach a new world record.

And as dancers know, when they lose themselves in the dance, synchrony is also a portal to another dimension – to that sacred place where the boundaries of the self grow fuzzy, and time seems not to exist. I've felt it as a member of a community choir, when the sharp and flat tones of amateur voices combine into a perfection that few of the singers involved could attain on their own. (The extraordinary psychological benefits of choral singing, one 2005 study drily concluded, are not reduced 'when the vocal instrument is of mediocre quality'.) For that matter, I've felt it in settings that are even more mundane – working my monthly shift at the food cooperative, for example, slinging boxes of carrots and broccoli onto the conveyor belt, in time with other workers I barely know but with whom, for a few hours, I share a bond that feels deeper than the one I have with some of my real friends. For a while, it's as if we're participating in the communal rhythms of a monastery, in which the synchronised hours of prayer and labour impart coherence and a sense of shared purpose to the day.

Something mysterious is at work in such moments – and there may be no better proof of how potent it is than the fact that it can be harnessed for dangerous and indeed fatal ends. From the point of view of military commanders, after all, the chief benefit of synchrony among soldiers isn't that they'll march for longer distances. It's that once they feel they belong to something greater than themselves, they'll be more willing to lay down their life for their unit. Midway through a rehearsal of Handel's *Messiah* in a high-ceilinged church, it becomes just about possible for an amateur singer to imagine how a person might enter that state of mind. The world 'doesn't open up into a million shimmering dimensions of hope and possibility when I sing alone', the writer and choir member Stacy Horn observes. That happens only 'when I'm surrounded by my fellow choristers, and all the different sounds we're making combine to leave us thrumming in harmony – lit up together like fireflies flashing in synchrony by whatever masterpiece is currently racing through our brains, bodies, and hearts'.

The Freedom to Never See Your Friends

The question is, What kind of freedom do we really want when it comes to time? On the one hand, there's the culturally celebrated goal of individual time sovereignty – the freedom to set your own schedule, to make your own choices, to be *free from* other people's intrusions into your precious four thousand weeks. On the other hand, there's the profound sense of meaning that comes from being willing to fall in with the rhythms of the rest of the world: to be *free to* engage in all the worthwhile collaborative endeavours that require at least some sacrifice of your sole control over what you do and when. Strategies for achieving the first kind of freedom are the sort of thing that fills books of productivity advice: ideal morning routines, strict personal schedules, and tactics for limiting how long you spend answering email each day, plus homilies on the importance of 'learning to say no' – all of them functioning as bulwarks against the risk that other people might exert too much influence over how your time gets used. And undoubtedly these have a role to play: we do need to set firm boundaries so that bullying bosses, exploitative employment arrangements, narcissistic spouses or a guilty tendency towards people-pleasing don't end up dictating the course of every day.

And yet the trouble with this kind of individualist freedom, as Judith Shulevitz points out, is that a society in thrall to it, as ours is, ends up desynchronising itself – imposing upon itself something surprisingly similar, in its results, to the disastrous Soviet experiment with a staggered five-day week. We live less and less of our lives in the same temporal grooves as one another. The unbridled reign of this individualist ethos, fuelled by the demands of the market economy, has overwhelmed our traditional ways of organising time, meaning that the hours in which we rest, work and socialise are becoming ever more uncoordinated. It's harder than ever to find time for a leisurely family dinner, a spontaneous visit to friends, or any collective project – nurturing a community garden, playing in an amateur rock band – that takes place in a setting other than the workplace.

For the least privileged, the dominance of this kind of freedom translates into no freedom at all: it means unpredictable gig-economy jobs and 'on-demand scheduling', in which the giant retailer you work for might call you into work at any moment, its labour needs calculated algorithmically from hour to hour based on sales volume – making it all but impossible to plan childcare or essential visits to the doctor, let alone a night out with friends. But even for those of us who genuinely do have much more personal control over when we work than previous generations ever did, the result is that work seeps through life like water, filling every cranny with more to-dos, a phenomenon that seemed to only intensify during the coronavirus lockdown. It starts to feel as though you, your spouse and your closest friends have all been assigned to different colour-coded Soviet work groups. The reason it's so hard for my wife and me to find an hour in the week for a serious conversation, or for me and my three closest friends to meet for a beer, isn't usually that we 'don't have the time', in the strict sense of that phrase, though that's what we may tell ourselves. It's that we do have the time – but that there's almost no likelihood of it being the same portion of time for everyone involved. Free to pursue our own entirely personal schedules, yet still yoked to our jobs, we've constructed lives that can't be made to mesh.

All this comes with political implications, too, because grassroots politics – the world of meetings, rallies, protests and canvassing – are among the most important coordinated activities that a desynchronised population finds it difficult to get round to doing. The result is a vacuum of collective action, which gets filled by autocratic leaders, who thrive on the mass support of people who are otherwise disconnected – alienated from one another, stuck at home on the couch, a captive audience for televised propaganda. 'Totalitarian movements are mass organizations of atomized, isolated individuals,' wrote Hannah Arendt in *The Origins of Totalitarianism*. It's in the interests of an autocrat that the only real bond among his supporters should be their support for him. On those occasions when synchronised action does pierce through the isolation, as during the worldwide demonstrations that followed the killing of George Floyd by Minneapolis police in 2020, it's not unusual to hear protesters describe experiences that call to mind William McNeill's 'strange sense of personal enlargement' – a feeling of time thickening and intensifying, tinged with a kind of ecstasy.

Like our other troubles with time, our loss of synchrony obviously can't be solved exclusively at the level of the individual or the family. (Good luck persuading everyone in your neighbourhood to take the same day off work each week.) But we do each get to decide whether to collaborate with the ethos of individual time sovereignty or to resist it. You *can* push your life a little further in the direction of the second, communal sort of freedom. For one thing, you can make the kinds of commitments that remove flexibility from your schedule in exchange for the rewards of community, by joining amateur choirs or sports teams, campaign groups or religious organisations. You can prioritise activities in the physical world over those in the digital one, where even collaborative activity ends up feeling curiously isolating. And if, like me, you possess the productivity geek's natural inclination towards control-freakery when it comes to your time, you can experiment with what it feels like to *not* try to exert an iron grip on your timetable: to sometimes let the rhythms of family life and friendships and collective action take precedence over your perfect morning routine or your system for scheduling your week. You can grasp the truth that power over your time isn't something best hoarded entirely for yourself: that your time can be too much your own.

13.

Cosmic Insignificance Therapy

The Jungian psychotherapist James Hollis recalls the experience of one of his patients, a successful vice president of a medical instruments company, who was flying over the American Midwest on a business trip, reading a book, when she was accosted by a thought: 'I hate my life.' A malaise that had been growing in her for years had crystallised in the understanding that she was spending her days in a way that no longer felt as if it had any meaning. The relish she'd had for her work had drained away; the rewards she'd been pursuing seemed worthless; and now life was a matter of going through the motions, in the fading hope that it somehow all might yet pay off in future happiness.

Perhaps you know how she felt. Not everyone has this kind of sudden epiphany, but many of us know what it is to suspect that there might be richer, fuller, juicier things we could be doing with our four thousand weeks – even when what we're currently doing with them looks, from the outside, like the definition of success. Or maybe you're familiar with the experience of returning to your daily routines, following an unusually satisfying weekend in nature or with old friends, and being struck by the thought that more of life should feel that way – that it wouldn't be unreasonable to expect the deeply engrossing parts to be more than rare exceptions. The modern world is especially lacking in good responses to such feelings: religion no longer provides the universal ready-made sense of purpose it once did, while consumerism misleads us into seeking meaning where it can't be found. But the sentiment itself is an ancient one. The writer of the book of Ecclesiastes, among many other people, would instantly have recognised the suffering of Hollis's patient: 'Then I considered all that my hands had done, and the toil I had spent in doing it, and behold, all was vanity and a striving after wind, and there was nothing to be gained under the sun.'

It's deeply unsettling to find yourself doubting the point of what you're doing with your life. But it isn't actually a bad thing, because it demonstrates that an inner shift has already occurred. You couldn't entertain such doubts in the first place if you weren't already occupying a new vantage point on your life – one from which you'd already begun to face the reality that you can't depend on fulfilment arriving at some distant point in the future, once you've got your life in order, or met the world's criteria for success, and that instead the matter needs addressing now. To realise midway through a business trip that you hate your life is already to have taken the first step into one you don't hate – because it means you've grasped the fact that *these* are the weeks that are going to have to be spent doing something worthwhile, if your finite life is to mean anything at all. This is a perspective from which you can finally ask the most fundamental question of time management: what would it mean to spend the only time you ever get in a way that truly feels as though you are making it count?

The Great Pause

Sometimes this perceptual jolt affects a whole society at once. I wrote the first draft of this chapter under lockdown in New York City, during the coronavirus pandemic, when, amid the grief and anxiety, it became normal to hear people express a sort of bitter-sweet gratitude for what they were experiencing: that even though they were furloughed and losing sleep about the rent, it was a genuine joy to see more of their children, or to rediscover the pleasures of planting flowers or baking bread. The enforced pause in work, school and socialising put on hold numerous assumptions about how we had to spend our time. It turned out, for example, that many people could perform their jobs adequately without an hour-long commute to a dreary office, or remaining at a desk until 6.30 p.m. solely in order to appear hard-working. It also turned out that most of the restaurant meals and takeaway coffees I'd grown accustomed to consuming, presumably on the grounds that they enhanced my life, could be forsworn with no feeling of loss (a double-edged revelation, given how many jobs depended on providing them). And it became clear – from the ritual applauding of emergency workers, shopping trips undertaken for housebound neighbours, and many other acts of generosity – that people cared about one another far more than we'd assumed. It was just that before the virus, apparently, we hadn't had the time to show it.

Things hadn't changed for the better, obviously. But alongside the devastation that it wrought, the virus changed *us* for the better, at least temporarily, and at least in certain respects: it helped us perceive more clearly what our pre-lockdown days had been lacking and the trade-offs we'd been making, willingly or otherwise – for example, by pursuing work lives that left no time for neighbourliness. A New York writer and director named Julio Vincent Gambuto captured this sense of what I found myself starting to think of as 'possibility shock' – the startling understanding that things could be different, on a grand scale, if only we collectively wanted that enough. 'What the trauma has shown us,' Gambuto wrote, 'cannot be unseen. A carless Los Angeles has clear blue skies, as pollution has simply stopped. In a quiet New York, you can hear the birds chirp in the middle of Madison Avenue. Coyotes have been spotted on the Golden Gate Bridge. These are the postcard images of what the world might be like if we could find a way to have a less deadly effect on the planet.' Of course, the crisis also revealed underfunded healthcare systems, venal politicians, deep racial inequities and chronic economic insecurity. But these, too, contributed to the feeling that now we were seeing what actually mattered, what demanded our attention – and that on some level we'd known it all along.

When lockdown ended, Gambuto warned, corporations and governments would conspire to make us forget the possibilities we'd glimpsed, by means of shiny new products and services and distracting culture wars; and we'd be so desperate to return to normality that we'd be tempted to comply. Instead, though, we could hold on to the sense of strangeness and make new choices about how we used the hours of our lives:

> What happened is inexplicably incredible. It's the greatest gift ever unwrapped. Not the deaths, not the virus, but The Great Pause . . . Please don't recoil from the bright light beaming through the window. I know it hurts your eyes. It hurts mine, too. But the curtain is wide open . . . The Great American Return to Normal is coming . . . [but] I beg of you: take a deep breath, ignore the deafening noise, and think deeply about what you want to put back into your life. This is our chance to define a new version of normal, a rare and truly sacred (yes, sacred) opportunity to get rid of the bullshit and to only bring back what works for us, what makes our lives richer, what makes our kids happier, what makes us truly proud.

The hazard in any such discussion of 'what matters most' in life, though, is that it tends to give rise to a kind of paralysing grandiosity. It starts to feel as though it's your duty to find something truly consequential to do with your time – to quit your office job to become an aid worker or start a space flight company – or else, if you're in no position to make such a grand gesture, to conclude that a deeply meaningful life isn't an option for you. On the level of politics and social change, it becomes tempting to conclude that only the most revolutionary, world-transforming causes are worth fighting for – that it would be meaningless to spend your time, say, caring for an elderly relative with dementia or volunteering at the local community garden while the problems of global warming and income inequality remain unsolved. Among New Age types, this same grandiosity takes the form of the belief that each of us has some cosmically significant Life Purpose, which the universe is longing for us to uncover and then to fulfil.

Which is why it's useful to begin this last stage of our journey with a blunt but unexpectedly liberating truth: that what you do with your life doesn't matter all that much – and when it comes to how you're using your finite time, the universe absolutely could not care less.

A Modestly Meaningful Life

The late philosopher Bryan Magee liked to make the following arresting point. Human civilisation is about six thousand years old, and we're in the habit of thinking of this as a staggeringly long time: a vast duration across which empires rose and fell, and historical periods to which we give labels such as 'classical antiquity' or 'the Middle Ages' succeeded each other in 'only-just-moving time – time moving in the sort of way a glacier moves'. But now consider the matter a different way. In every generation, even back when life expectancy was much shorter than it is today, there were always at least a few people who lived to the age of one hundred (or 5,200 weeks). And when each of *those* people was born, there must have been a few other people alive at the time who had already reached the age of one hundred themselves. So it's possible to visualise a chain of centenarian lifespans, stretching all the way back through history, with no spaces in between them: specific people who really lived, and each of whom we could name, if only the historical record were good enough.

Now for the arresting part: by this measure, the golden age of the Egyptian pharaohs – an era that strikes most of us as impossibly remote from our own – took place a scant thirty-five lifetimes ago. Jesus was born about twenty lifetimes ago, and the Renaissance happened seven lifetimes back. A paltry five centenarian lifetimes ago, Henry VIII sat on the English throne. Five! As Magee observed, the number of lives you'd need in order to span the whole of civilisation, sixty, was 'the number of friends I squeeze into my living room when I have a drinks party'. From this perspective, human history hasn't unfolded glacially but in the blink of an eye. And it follows, of course, that your own life will have been a minuscule little flicker of near-nothingness in the scheme of things: the merest pinpoint, with two incomprehensibly vast tracts of time, the past and future of the cosmos as a whole, stretching off into the distance on either side.

It's natural to find such thoughts terrifying. To contemplate 'the massive indifference of the universe', writes Richard Holloway, the former Bishop of Edinburgh, can feel 'as disorienting as being lost in a dense wood, or as frightening as falling overboard into the sea with no-one to know we have gone'. But there's another angle from which it's oddly consoling. You might think of it as 'cosmic insignificance therapy': when things all seem too much, what better solace than a reminder that they are, provided you're willing to zoom out a bit, indistinguishable from nothing at all? The anxieties that clutter the average life – relationship troubles, status rivalries, money worries – shrink instantly down to irrelevance. So do pandemics and presidencies, for that matter: the cosmos carries on regardless, calm and imperturbable. Or to quote the title of a book I once reviewed: *The Universe Doesn't Give a Flying Fuck About You*. To remember how little you matter, on a cosmic timescale, can feel like putting down a heavy burden that most of us didn't realise we were carrying in the first place.

This sense of relief is worth examining a little more closely, though, because it draws attention to the fact that the rest of the time, most of us *do* go around thinking of ourselves as fairly central to the unfolding of the universe; if we didn't, it wouldn't be any relief to be reminded that in reality this isn't the case. Nor is this a phenomenon confined to megalomaniacs or pathological narcissists, but something much more fundamental to being human: it's the understandable tendency to judge everything from the perspective you occupy, so that the few thousand weeks for which *you* happen to be around inevitably come to feel like the linchpin of history, to which all prior time was always leading up. These self-centred judgements are part of what psychologists call the 'egocentricity bias', and they make good sense from an evolutionary standpoint. If you had a more realistic sense of your own sheer irrelevance, considered on the timescale of the universe, you'd probably be less motivated to struggle to survive, and thereby to propagate your genes.

You might imagine, moreover, that living with such an unrealistic sense of your own historical importance would make life feel more meaningful, by investing your every action with a feeling of cosmic significance, however unwarranted. But what actually happens is that this overvaluing of your existence gives rise to an unrealistic definition of what it would mean to use your finite time well. It sets the bar much too high. It suggests that in order to count as having been well spent', your life needs to involve deeply impressive accomplishments, or that it should have a lasting impact on future generations – or at the very least that it must, in the words of the philosopher Iddo Landau, 'transcend the common and the mundane'. Clearly, it can't just be ordinary: after all, if your life is as significant in the scheme of things as you tend to believe, how could you *not* feel obliged to do something truly remarkable with it?

This is the mindset of the Silicon Valley tycoon determined to 'put a dent in the universe', or the politician fixated on leaving a legacy, or the novelist who secretly thinks her work will count for nothing unless it reaches the heights, and the public acclaim, of Leo Tolstoy's. Less obviously, though, it is also the implicit outlook of those who glumly conclude that their life is ultimately meaningless, and that they'd better stop expecting it to feel otherwise. What they really mean is that they've adopted a standard of meaningfulness to which virtually nobody could ever measure up. 'We do not disapprove of a chair because it cannot be used to boil water for a nice cup of tea,' Landau points out: a chair just isn't the kind of thing that ought to have the capacity to boil water, so it isn't a problem that it doesn't. And it is likewise 'implausible, for almost all people, to demand of themselves that they be a Michelangelo, a Mozart, or an Einstein . . . There have only been a few dozen such people in the entire history of humanity.' In other words, you almost certainly *won't* put a dent in the universe. Indeed, depending on the stringency of your criteria, even Steve Jobs, who coined that phrase, failed to leave such a dent. Perhaps the iPhone will be remembered for more generations than anything you or I will ever accomplish; but from a truly cosmic view, it will soon be forgotten, like everything else.

No wonder it comes as a relief to be reminded of your insignificance: it's the feeling of realising that you'd been holding yourself, all this time, to standards you couldn't reasonably be expected to meet. And this realisation isn't merely calming but liberating, because once you're no longer burdened by such an unrealistic definition of a 'life well spent', you're freed to consider the possibility that a far wider variety of things might qualify as meaningful ways to use your finite time. You're freed, too, to consider the possibility that many of the things you're *already* doing with it are more meaningful than you'd supposed – and that until now, you'd subconsciously been devaluing them, on the grounds that they weren't 'significant' enough.

From this new perspective, it becomes possible to see that preparing nutritious meals for your children might matter as much as anything could ever matter, even if you won't be winning any cooking awards; or that your novel's worth writing if it moves or entertains a handful of your contemporaries, even though you know you're no Tolstoy. Or that virtually any career might be a worthwhile way to spend a working life, if it makes things slightly better for those it serves. Furthermore, it means that if what we learn from the experience of the coronavirus pandemic is to become just a little more attuned to the needs of our neighbours, we'll have learned something valuable as a result of the 'Great Pause', no matter how far off the root-and-branch transformation of society remains.

Cosmic insignificance therapy is an invitation to face the truth about your irrelevance in the grand scheme of things. To embrace it, to whatever extent you can. (Isn't it hilarious, in hindsight, that you ever imagined things might be otherwise?) Truly doing justice to the astonishing gift of a few thousand weeks isn't a matter of resolving to 'do something remarkable' with them. In fact, it entails precisely the opposite: refusing to hold them to an abstract and overdemanding standard of remarkableness, against which they can only ever be found wanting, and taking them instead on their own terms, dropping back down from godlike fantasies of cosmic significance into the experience of life as it concretely, finitely – and often enough, marvellously – really is.

14.

The Human Disease

The fantasy behind so many of our time-related troubles is the one encapsulated in the title of the book I alluded to in the first chapter: *Master Your Time, Master Your Life*, by the time management guru Brian Tracy. The reason time feels like such a struggle is that we're constantly attempting to master it – to lever ourselves into a position of dominance and control over our unfolding lives so that we might finally feel safe and secure, and no longer so vulnerable to events.

For some of us, the struggle manifests as the attempt to become so productive and efficient that we never again have to experience the guilt of disappointing others, or worry about being fired for underperforming; or so that we might avoid facing the prospect of dying without having fulfilled our greatest ambitions. Other people hold off entirely from starting on important projects or embarking on intimate relationships in the first place because they can't bear the anxiety of having committed themselves to something that might or might not work out happily in practice. We waste our lives railing against traffic jams and toddlers for having the temerity to take the time they take, because they're blunt reminders of how little control we truly have over our schedules. And we chase the ultimate fantasy of time mastery – the desire, by the time we die, to have truly mattered in the cosmic scheme of things, as opposed to being instantly trampled underfoot by the advancing aeons.

This dream of somehow one day getting the upper hand in our relationship with time is the most forgivable of human delusions because the alternative is so unsettling. But unfortunately, it's the alternative that's true: the struggle is doomed to fail. Because your quantity of time is so limited, you'll never reach the commanding position of being able to handle every demand that might be thrown at you or pursue every ambition that feels important; you'll be obliged to make tough choices instead. And because you can't dictate, or even accurately predict, so much of what happens with the finite portion of time you *do* get, you'll never feel that you're securely in charge of events, immune from suffering, primed and ready for whatever comes down the pike.

The Provisional Life

But the deeper truth behind all this is to be found in Heidegger's mysterious suggestion that we don't *get* or *have* time at all – that instead we *are* time. We'll never get the upper hand in our relationship with the moments of our lives because we are nothing but those moments. To 'master' them would first entail getting outside of them, splitting off from them. But where would we go? 'Time is the substance I am made of,' writes Jorge Luis Borges. 'Time is a river that sweeps me along, but I am the river; it is a tiger which destroys me, but I am the tiger; it is a fire which consumes me, but I am the fire.' There's no scrambling up to the safety of the riverbank when the river is you. And so insecurity and vulnerability are the default state – because in each of the moments that you inescapably are, anything could happen, from an urgent email that scuppers your plans for the morning to a bereavement that shakes your world to its foundations.

A life spent focused on achieving security with respect to time, when in fact such security is unattainable, can only ever end up feeling provisional – as if the point of your having been born still lies in the future, just over the horizon, and your life in all its fullness can begin as soon as you've put it, in Arnold Bennett's phrase, 'into proper working order'. Once you've cleared the decks, you tell yourself; or once you've implemented a better system of personal organisation, or got your degree, or invested a sufficient number of years in honing your craft; or once you've found your soulmate or had kids, or once the kids have left home, or once the revolution comes and social justice is established – *that's* when you'll feel in control at last, you'll be able to relax a bit, and true meaningfulness will be found. Until then, life necessarily feels like a struggle: sometimes an exciting one, sometimes exhausting, but always in the service of some moment of truth that's still in the future. Writing in 1970, Marie-Louise von Franz, the Swiss psychologist and scholar of fairy tales, captured the other-worldly atmosphere of such an existence:

> There is a strange attitude and feeling that one is *not yet* in real life. For the time being one is doing this or that, but whether it is [a relationship with] a woman or a job, it is *not yet* what is really wanted, and there is always the fantasy that sometime in the future the real thing will come about . . . The one thing dreaded throughout by such a type of man is to be bound to anything whatever. There is a terrific fear of being pinned down, of entering space and time completely, and of being the unique human that one is.

'Entering space and time completely' – or even partially, which may be as far as any of us ever get – means admitting defeat. It means letting your illusions die. You have to accept that there will always be too much to do; that you can't avoid tough choices or make the world run at your preferred speed; that no experience, least of all close relationships with other human beings, can ever be guaranteed in advance to turn out painlessly and well – and that from a cosmic viewpoint, when it's all over, it won't have counted for very much anyway.

And in exchange for accepting all that? You get to actually *be* here. You get to have some real purchase on life. You get to spend your finite time focused on a few things that matter to you, in themselves, right now, in this moment. Maybe it's worth spelling out that none of this is an argument against long-term endeavours like marriage or parenting, building organisations or reforming political systems, and certainly not against tackling the climate crisis; these are among the things that matter most. But it's an argument that even those things can only ever matter now, in each moment of the work involved, whether or not they've yet reached what the rest of the world defines as fruition. Because now is all you ever get.

It's tempting to imagine that ending or at least easing the struggle with time might also make you *happy*, most or all of the time. But I've no reason to believe that's true. Our finite lives are filled with all the painful problems of finitude, from overfilled inboxes to death, and confronting them doesn't stop them from feeling like problems – or not exactly, anyway. The peace of mind on offer here is of a higher order: it lies in the recognition that being unable to escape from the problems of finitude is not, in itself, a problem. The human disease is often painful, but as the Zen teacher Charlotte Joko Beck puts it, it's only *unbearable* for as long as you're under the impression that there might be a cure. Accept the inevitability of the affliction, and freedom ensues: you can get on with living at last. The same realisation that struck me on that park bench in Brooklyn struck the French poet Christian Bobin, he recalls, at a similarly mundane moment: 'I was peeling a red apple from the garden when I suddenly understood that life would only ever give me a series of wonderfully insoluble problems. With that thought an ocean of profound peace entered my heart.'

Five Questions

To make this all a little more concrete, it may be useful to ask the following questions of your own life. It doesn't matter if answers aren't immediately forthcoming; the point, in Rainer Maria Rilke's famous phrase, is to 'live the questions'. Even to ask them with any sincerity is already to have begun to come to grips with the reality of your situation and to start to make the most of your finite time.

1. Where in your life or your work are you currently pursuing comfort, when what's called for is a little discomfort?

Pursuing the life projects that matter to you the most will almost always entail *not* feeling fully in control of your time, immune to the painful assaults of reality, or confident about the future. It means embarking on ventures that might fail, perhaps because you'll find you lacked sufficient talent; it means risking embarrassment, holding difficult conversations, disappointing others, and getting so deep into relationships that additional suffering – when bad things happen to those you care about – is all but guaranteed. And so we naturally tend to make decisions about our daily use of time that prioritise anxiety-avoidance instead. Procrastination, distraction, commitment-phobia, clearing the decks and taking on too many projects at once are all ways of trying to maintain the illusion that you're in charge of things. In a subtler way, so too is compulsive worrying, which offers its own gloomy but comforting sense that you're doing something constructive to try to stay in control.

James Hollis recommends asking of every significant decision in life: 'Does this choice diminish me, or enlarge me?' The question circumvents the urge to make decisions in the service of alleviating anxiety and instead helps you make contact with your deeper intentions for your time. If you're trying to decide whether to leave a given job or relationship, say, or to redouble your commitment to it, asking what would make you happiest is likely to lure you towards the most comfortable option, or else leave you paralysed by indecision. But you usually know, intuitively, whether remaining in a relationship or job would present the kind of challenges that will help you grow as a person (enlargement) or the kind that will cause your soul to shrivel with every passing week (diminishment). Choose uncomfortable enlargement over comfortable diminishment whenever you can.

2. Are you holding yourself to, and judging yourself by, standards of productivity or performance that are impossible to meet?

One common symptom of the fantasy of someday achieving total mastery over time is that we set ourselves inherently impossible targets for our use of it – targets that must always be postponed into the future, since they can never be met in the present. The truth is that it's impossible to become so efficient and organised that you could respond to a limitless number of incoming demands. It's usually equally impossible to spend what feels like 'enough time' on your work and with your children, and on socialising, travelling or engaging in political activism. But there's a deceptive feeling of comfort in believing that you're in the process of constructing such a life, which is due to come into being any day now.

What would you do differently with your time, today, if you knew in your bones that salvation was never coming – that your standards had been unreachable all along, and that you'll therefore never manage to make time for all you hoped you might? Perhaps you're tempted to object that yours is a special case, that in your particular situation you *do* need to pull off the impossible, timewise, in order to avert catastrophe. For example, maybe you're afraid you'll be fired and lose your income if you don't stay on top of your impossible workload. But this is a misunderstanding. If the level of performance you're demanding of yourself is genuinely impossible, then it's impossible, even if catastrophe looms – and facing this reality can only help.

There is a sort of cruelty, Iddo Landau points out, in holding yourself to standards nobody could ever reach (and which many of us would never dream of demanding of other people). The more humane approach is to drop such efforts as completely as you can. Let your impossible standards crash to the ground. Then pick a few meaningful tasks from the rubble and get started on them today.

3. In what ways have you yet to accept the fact that you are who you are, not the person you think you ought to be?

A closely related way to postpone the confrontation with finitude – with the anxiety-inducing truth that *this is it* – is to treat your present-day life as part of a journey towards becoming the kind of person you believe you *ought* to become, in the eyes of society, a religion, or your parents, whether or not they're still alive. Once you've earned your right to exist, you tell yourself, life will stop feeling so uncertain and out of control. In times of political and environmental crisis, this mindset often takes the form of the belief that nothing is truly worth doing with your time except addressing such emergencies head-on, round the clock – and that you're entirely correct to think of yourself as guilty and selfish for spending it on anything else.

This quest to justify your existence in the eyes of some outside authority can continue long into adulthood. But 'at a certain age', writes the psychotherapist Stephen Cope, 'it finally dawns on us that, shockingly, no one really *cares* what we're doing with our life. This is a most unsettling discovery to those of us who have lived someone else's life and eschewed our own: no one really cares except us.' The attempt to attain security by justifying your existence, it turns out, was both futile and unnecessary all along. Futile because life will always feel uncertain and out of your control. And unnecessary because, in consequence, there's no point in waiting to live until you've achieved validation from someone or something else. Peace of mind, and an exhilarating sense of freedom, comes not from achieving the validation but from yielding to the reality that it wouldn't bring security if you got it.

I'm convinced, in any case, that it is from this position of *not* feeling as though you need to earn your weeks on the planet that you can do the most genuine good with them.

Once you no longer feel the stifling pressure to become a particular kind of person, you can confront the personality, the strengths and weaknesses, the talents and enthusiasms you find yourself with, here and now, and follow where they lead. Perhaps your particular contribution to a world facing multiple crises isn't primarily to spend your time pursuing activism, or seeking electoral office, but on caring for an elderly relative, or making music, or working as a pastry chef, like my brother-in-law, a strapping South African you might mistake for a rugby player but whose work involves concocting intricate structures of spun sugar and butter icing that detonate small explosions of joy in their recipients. The Buddhist teacher Susan Piver points out that it can be surprisingly radical and discomfiting, for many of us, to ask how we'd *enjoy* spending our time. But at the very least, you shouldn't rule out the possibility that the answer to that question is an indication of how you might use your time best.

4. In which areas of life are you still holding back until you feel like you know what you're doing?

It's easy to spend years treating your life as a dress rehearsal on the rationale that what you're doing, for the time being, is acquiring the skills and experience that will permit you to assume authoritative control of things later on. But I sometimes think of my journey through adulthood to date as one of incrementally discovering the truth that there is no institution, no walk of life, in which everyone isn't just winging it, all the time. Growing up, I assumed that the newspaper on the breakfast table must be assembled by people who truly knew what they were doing; then I got a job at a newspaper. Unconsciously, I transferred my assumptions of competence elsewhere, including to people who worked in government. But then I got to know a few people who did – and who would admit, after a couple of drinks, that their jobs involved staggering from crisis to crisis, inventing plausible-sounding policies in the backs of cars en route to the press conferences at which those policies had to be announced. Even then, I found myself assuming that this might all be explained as a manifestation of the perverse pride that British people sometimes take in being shamblingly mediocre. Then I moved to America – where, it turns out, everyone is winging it, too. Political developments in the years since have only made it clearer that the people 'in charge' have no more command over world events than the rest of us do.

It's alarming to face the prospect that you might *never* truly feel as though you know what you're doing, in work, marriage, parenting or anything else. But it's liberating, too, because it removes a central reason for feeling self-conscious or inhibited about your performance in those domains in the present moment: if the feeling of total authority is never going to arrive, you might as well not wait any longer to give such activities your all – to put bold plans into practice, to stop erring on the side of caution. It is even more liberating to reflect that everyone else is in the same boat, whether they're aware of it or not.

5. How would you spend your days differently if you didn't care so much about seeing your actions reach fruition?

A final common manifestation of the desire for time mastery arises from the unspoken assumption described in chapter 8 as the causal catastrophe: the idea that the true value of how we spend our time is always and only to be judged by the results. It follows naturally enough from this outlook that you should focus your time on those activities for which you expect to be around to see the results. But in his documentary *A Life's Work*, the director David Licata profiles people who took another path, dedicating their lives to projects that almost certainly won't be completed within their lifetimes – like the father-and-son team attempting to catalogue every tree in the world's remaining ancient forests, and the astronomer scouring radio waves for signs of extraterrestrial life from her desk at the SETI Institute in California. All have the shining eyes of people who know they're doing things that matter, and who relish their work precisely because they don't need to try to convince themselves that their own contributions will prove decisive or reach fruition while they're still alive.

Yet there is a sense in which all work – including the work of parenting, community-building and everything else – has this quality of not being completable within our own lifetimes. All such activities always belong to a far bigger temporal context, with an ultimate value that will only be measurable long after we're gone (or perhaps never, since time stretches on indefinitely). And so it's worth asking: what actions – what acts of generosity or care for the world, what ambitious schemes or investments in the distant future – might it be meaningful to undertake today, if you could come to terms with never seeing the results? We're all in the position of medieval stonemasons, adding a few more bricks to a cathedral whose completion we know we'll never see. The cathedral's still worth building, all the same.

The Next Most Necessary Thing

On 15 December 1933, Carl Jung wrote a reply to a correspondent, Frau V., responding to several questions on the proper conduct of life, and his answer is a good one with which to end this book. 'Dear Frau V.,' Jung began, 'Your questions are unanswerable, because you want to know how to live. One lives as one can. There is no single, definite way . . . If that's what you want, you had best join the Catholic Church, where they tell you what's what.' By contrast, the individual path 'is the way you make for yourself, which is never prescribed, which you do not know in advance, and which simply comes into being itself when you put one foot in front of the other'. His sole advice for walking such a path was to 'quietly do the next and most necessary thing. So long as you think you don't yet know what that is, you still have too much money to spend in useless speculation. But if you do with conviction the next and most necessary thing, you are always doing something meaningful and intended by fate.' A modified version of this insight, 'Do the next right thing', has since become a slogan favoured among members of Alcoholics Anonymous, as a way to proceed sanely through moments of acute crisis. But really, the 'next and most necessary thing' is all that any of us can ever aspire to do in any moment. And we must do it despite not having any objective way to be sure what the right course of action even is.

Fortunately, precisely because that's all you can do, it's also all that you ever *have* to do. If you can face the truth about time in this way – if you can step more fully into the condition of being a limited human – you will reach the greatest heights of productivity, accomplishment, service and fulfilment that were ever in the cards for you to begin with. And the life you will see incrementally taking shape, in the rear-view mirror, will be one that meets the only definitive measure of what it means to have used your weeks well: not how many people you helped, or how much you got done; but that working within the limits of your moment in history, and your finite time and talents, you actually got round to doing – and made life more luminous for the rest of us by doing – whatever magnificent task or weird little thing it was that you came here for.

Afterword: Beyond Hope

Except there's a problem: everything is screwed. Perhaps you've noticed.

A time traveller from an ancient Hindu civilisation would have no difficulty recognising our era as part of the Kali Yuga, that phase in the cycle of history when, according to Hindu mythology, everything starts to unravel: governments crumble, the environment collapses and strange weather events proliferate, refugees pour across borders, and diseases and dubious ideologies spread across the world. (Much of that comes almost verbatim from the Mahabharata, the 2,000-year-old Sanskrit epic, so its resemblance to my Twitter timeline is either coincidental or extremely sinister.) It is true, as more upbeat commentators like to remind us, that people have always believed they were living in the end times, and that much of the news these days is really rather good: infant mortality, absolute poverty and global inequality are all falling rapidly, while literacy is rising, and you're less likely than ever to get killed in a war. Still, those 34°C days in the Arctic are real, too, as is the coronavirus pandemic, the epic wildfires, the dinghies overloaded with desperate migrants. To put things as mildly as possible, it's hard to remain entirely confident that everything will turn out fine.

Why focus on time management in an era like this? It might seem like the height of irrelevance. But as I've sought to make clear, I think that's mainly just a consequence of the blinkered focus of most conventional time management advice. Broaden your perspective a bit, and it's obvious that in periods of anxiety and darkness, questions of time use take on fresh urgency: our success or failure in responding to the challenges we face will turn entirely on how we use the hours available in the day. The phrase 'time management' might seem to render the whole thing rather mundane. But then a mundane life – in the sense of the one that's unfolding here now, in this very moment – is all that we have to work with.

People sometimes ask Derrick Jensen, the environmentalist who co-founded the radical group Deep Green Resistance, how he manages to stay hopeful when everything seems so grim. But he tells them he doesn't – and that he thinks that's a good thing. Hope is supposed to be 'our beacon in the dark', Jensen notes. But in reality, it's a curse. To *hope* for a given outcome is to place your faith in something outside yourself, and outside the current moment – the government, for example, or God, or the next generation of activists, or just 'the future' – to make things all right in the end. As the American Buddhist nun Pema Chödrön says, it means relating to life as if 'there's always going to be a babysitter available when we need one'. And sometimes that attitude can be justified: if I go into hospital for surgery, for instance, I do simply have to hope that the surgeon knows what he's doing, because no contribution I can make is likely to make much difference. But the rest of the time, it means disavowing your own capacity to change things – which in the context of Jensen's field, environmental activism, means surrendering your power to the very forces you were supposed to be fighting.

'Many people say they hope the dominant culture stops destroying the world,' as Jensen puts it, but by saying that, 'they've assumed the destruction will continue, at least in the short term, and they've stepped away from their own ability to participate in stopping it'. To give up hope, by contrast, is to reinhabit the power that you actually have. At that point, Jensen goes on, 'we no longer have to "hope" at all. We simply do the work. We make sure salmon survive. We make sure prairie dogs survive. We make sure grizzlies survive . . . When we stop hoping that the awful situation we're in will somehow resolve itself, when we stop hoping the situation will somehow not get worse, then we are finally free – truly free – to honestly start working to resolve it.'

You could think of this book as an extended argument for the empowering potential of giving up hope. Embracing your limits means giving up hope that with the right techniques, and a bit more effort, you'd be able to meet other people's limitless demands, realise your every ambition, excel in every role, or give every good cause or humanitarian crisis the attention it seems like it deserves. It means giving up hope of ever feeling totally in control, or certain that acutely painful experiences aren't coming your way. And it means giving up, as far as possible, the master hope that lurks beneath all this, the hope that somehow this isn't really *it* – that this is just a dress rehearsal, and that one day you'll feel truly confident that you have what it takes.

The key to what Chödrön calls 'getting the hang of hopelessness' lies in seeing that things *aren't* going to be okay. Indeed, they're already not okay – on a planetary level or an individual one. The Arctic ice is already melting. The pandemic has already killed millions, and crashed the economy. The question of how ill-qualified you can be for the American presidency yet still end up in the White House has already been definitively answered. Thousands of species are already gone. As one woman said, in a *New York Times* article about city-dwellers learning how to survive in the woods on deer meat and berries: 'People say, "Oh, when the apocalypse comes . . ." What are you talking about? It's here.' The world is already broken. And what's true of the state of civilisation is equally true of your life: it was always already the case that you would never experience a life of perfect accomplishment or security. And your four thousand weeks have always been running out.

It's a revelation, though: when you begin to internalise all this even just a bit, the result is not despair, but an energising surge of motivation. You come to see that the terrible eventuality against which you'd spent your life subliminally tensing your muscles, because it would be too appalling to experience, has already happened – and yet here you are, still alive, at least for the time being. 'Abandoning hope is an affirmation, the beginning of the beginning,' Chödrön says. You realise that you never really needed the feeling of complete security you'd previously felt so desperate to attain. This is a liberation. Once you no longer need to convince yourself that the world isn't filled with uncertainty and tragedy, you're free to focus on doing what you can to help. And once you no longer need to convince yourself that you'll do everything that needs doing, you're free to focus on doing a few things that count.

Although another way of making the point that giving up hope doesn't kill you, as Jensen points out, is that in a certain sense it does kill you. It kills the fear-driven, control-chasing, ego-dominated version of you – the one who cares intensely about what others think of you, about not disappointing anyone or stepping too far out of line, in case the people in charge find some way to punish you for it later. You find, Jensen writes, that 'the civilised you died. The manufactured, fabricated, stamped, molded you died. The victim died.' And the 'you' that remains is more alive than before. More ready for action, but also more joyful, because it turns out that when you're open enough to confront how things really are, you're open enough to let all the good things in more fully, too, on their own terms, instead of trying to use them to bolster your need to know that everything will turn out fine. You get to appreciate life in the droll spirit of George Orwell, on a stroll through a war-dazed London in early 1946, watching kestrels darting above the grim shadows of the gasworks, and tadpoles dancing in roadside streams, and later writing of the experience: 'Spring is here, even in London N1, and they can't stop you enjoying it.'

The average human lifespan is absurdly, terrifyingly, insultingly short. But that isn't a reason for unremitting despair, or for living in an anxiety-fuelled panic about making the most of your limited time. It's a cause for relief. You get to give up on something that was always impossible – the quest to become the optimised, infinitely capable, emotionally invincible, fully independent person you're officially supposed to be. Then you get to roll up your sleeves and start work on what's gloriously possible instead.

Appendix: Ten Tools for Embracing Your Finitude

In this book, I've made the case for embracing the truth about your limited time and limited control over that time – not simply because it's the truth, so you might as well face it, but because it's actively empowering to do so. By stepping more fully into reality as it actually is, you get to accomplish more of what matters, and feel more fulfilled about it. Here, in addition to the suggestions throughout the text, are ten further techniques for implementing this limit-embracing philosophy in daily life.

1. Adopt a 'fixed volume' approach to productivity.

Much advice on getting things done implicitly promises that it'll help you get *everything* important done – but that's impossible, and struggling to get there will only make you busier (see chapter 2). It's better to begin from the assumption that tough choices are inevitable and to focus on making them consciously and well. Any strategy for limiting your work in progress will help here (page 75), but perhaps the simplest is to **keep two to-do lists, one 'open' and one 'closed'.** The open list is for everything that's on your plate and will doubtless be nightmarishly long. Fortunately, it's not your job to tackle it: instead, feed tasks from the open list to the closed one – that is, a list with a fixed number of entries, ten at most. The rule is that you can't add a new task until one's completed. (You may also require a third list, for tasks that are 'on hold' until someone else gets back to you.) You'll never get through all the tasks on the open list – but you were never going to in any case, and at least this way you'll complete plenty of things you genuinely care about.

A complementary strategy is to establish **predetermined time boundaries for your daily work**. To whatever extent your job situation permits, decide in advance how much time you'll dedicate to work – you might resolve to start by 8.30 a.m., and finish no later than 5.30 p.m., say – then make all other time-related decisions in light of those predetermined limits. 'You could fill any arbitrary number of hours with what feels to be productive work,' writes Cal Newport, who explores this approach in his book *Deep Work*. But if your primary goal is to do what's required in order to be finished by 5.30, you'll be aware of the constraints on your time, and more motivated to use it wisely.

2. Serialise, serialise, serialise.

Following the same logic, **focus on one big project at a time** (or at most, one work project and one non-work project) and see it to completion before moving on to what's next. It's alluring to try to alleviate the anxiety of having too many responsibilities or ambitions by getting started on them all at once, but you'll make little progress that way; instead, train yourself to get incrementally better at *tolerating* that anxiety, by consciously postponing everything you possibly can, except for one thing. Soon, the satisfaction of completing important projects will make the anxiety seem worthwhile – and since you'll be finishing more and more of them, you'll have less to feel anxious about anyway. Naturally, it won't be possible to postpone absolutely everything – you can't stop paying the bills, or answering emails, or taking the kids to school – but this approach will ensure that the only tasks you don't postpone, while addressing your current handful of big projects, are the truly essential ones, rather than those you're dipping into solely to quell your anxiety.

3. Decide in advance what to fail at.

You'll inevitably end up underachieving at something, simply because your time and energy are finite. But the great benefit of **strategic underachievement** – that is, nominating in advance whole areas of life in which you won't expect excellence of yourself – is that you focus that time and energy more effectively. Nor will you be dismayed when you fail at what you'd planned to fail at all along. 'When you can't do it all, you feel ashamed and give up,' notes the author Jon Acuff, but when you 'decide in advance what things you're going to bomb . . . you remove the sting of shame'. A poorly kept lawn or a cluttered kitchen are less troubling when you've preselected 'lawn care' or 'kitchen tidiness' as goals to which you'll devote zero energy.

As with serialising your projects, there'll be plenty you can't choose to 'bomb' if you're to earn a living, stay healthy, be a decent partner and parent, and so forth. But even in these essential domains, there's scope to **fail on a cyclical basis**: to aim to do the bare minimum at work for the next two months, for example, while you focus on your children, or let your fitness goals temporarily lapse while you apply yourself to election canvassing. Then switch your energies to whatever you were neglecting. To live this way is to replace the high-pressure quest for 'work–life balance' with a conscious form of *im*balance, backed by your confidence that the roles in which you're underperforming right now will get their moment in the spotlight soon.

4. Focus on what you've already completed, not just on what's left to complete.

Since the quest to get everything done is interminable by definition (pages 39–44), it's easy to grow despondent and self-reproachful: you can't feel good about yourself until it's all finished – but it's never finished, so you never get to feel good about yourself. Part of the problem here is an unhelpful assumption that you begin each morning in a sort of 'productivity debt', which you must struggle to pay off through hard work, in the hope that you might reach a zero balance by the evening. As a counterstrategy, **keep a 'done list',** which starts empty first thing in the morning, and which you then gradually fill with whatever you accomplish through the day. Each entry is another cheering reminder that you *could*, after all, have spent the day doing nothing remotely constructive – and look what you did instead! (If you're in a serious psychological rut, lower the bar for what gets to count as an accomplishment: nobody else need ever know that you added 'brushed teeth' or 'made coffee' to the list.) Yet this is no mere exercise in consolation: there's good evidence for the motivating power of 'small wins', so the likely consequence of commemorating your minor achievements in this fashion is that you'll achieve more of them, and less-minor ones besides.

5. Consolidate your caring.

Social media is a giant machine for getting you to spend your time caring about the wrong things (pages 94–99), but for the same reason, it's also a machine for getting you to care about *too many* things, even if they're each indisputably worthwhile. We're exposed, these days, to an unending stream of atrocities and injustice – each of which might have a legitimate claim on our time and our charitable donations, but which in aggregate are more than any one human could ever effectively address. (Worse, the logic of the attention economy obliges campaigners to present whatever crisis they're addressing as uniquely urgent. No modern fundraising organisation would dream of describing its cause as the fourth- or fifth-most important of the day.)

Once you grasp the mechanisms operating here, it becomes easier to **consciously pick your battles in charity, activism and politics**: to decide that *your* spare time, for the next couple of years, will be spent lobbying for prison reform and helping at a local food pantry – not because fires in the Amazon or the fate of refugees don't matter, but because you understand that to make a difference, you must focus your finite capacity for care.

6. Embrace boring and single-purpose technology.

Digital distractions are so seductive because they seem to offer the chance of escape to a realm where painful human limitations don't apply: you need never feel bored or constrained in your freedom of action, which isn't the case when it comes to work that matters (pages 105–109). You can combat this problem by making your devices as boring as possible – first by removing social media apps, even email if you dare, and then by **switching the screen from colour to greyscale**. (At the time of writing, on the iPhone, this option can be found under Settings > Accessibility > Accessibility Shortcut > Colour Filters.) 'After going to grayscale, I'm not a different person all of a sudden, but I feel more in control of my phone, which now looks like a tool rather than a toy,' the technology journalist Nellie Bowles writes in the *New York Times*. Meanwhile, as far as possible, **choose devices with only one purpose**, such as the Kindle e-reader, on which it's tedious and awkward to do anything but read. If streaming music and social media lurk only a click or swipe away, they'll prove impossible to resist when the first twinge of boredom or difficulty arises in the activity on which you're attempting to focus.

7. Seek out novelty in the mundane.

It turns out that there may be a way to lessen, or even reverse, the dispiriting manner in which time seems to speed up as we age, so that the fewer weeks we have left, the faster we seem to lose them (page 7). The likeliest explanation for this phenomenon is that our brains encode the passage of years on the basis of how much information we process in any given interval. Childhood involves plentiful novel experiences, so we remember it as having lasted forever; but as we get older, life gets routinised – we stick to the same few places of residence, the same few relationships and jobs – and the novelty tapers off. 'As each passing year converts . . . experience into automatic routine,' wrote William James, soon 'the days and the weeks smooth themselves out in recollection to contentless units, and the years grow hollow and collapse'.

The standard advice for counteracting this is to cram your life with novel experiences, and this does work. But it's liable to worsen another problem, 'existential overwhelm' (pages 44–47). Moreover, it's impractical: if you have a job or children, much of life will necessarily be somewhat routine, and opportunities for exotic travel may be limited. An alternative, Shinzen Young explains, is to **pay more attention to every moment, however mundane**: to find novelty not by doing radically different things but by plunging more deeply into the life you already have. Experience life with twice the usual intensity, and 'your experience of life would be *twice as full* as it currently is' – and any period of life would be remembered as having lasted twice as long. Meditation helps here. But so does going on unplanned walks to see where they lead you, using a different route to get to work, taking up photography or birdwatching or nature drawing or keeping a journal, playing I Spy with a child: anything that draws your attention more fully into what you're doing in the present.

8. Be a 'researcher' in relationships.

The desire to feel securely in control of how our time unfolds causes numerous problems in relationships, where it manifests not just in overtly 'controlling' behaviour but in commitment-phobia, the inability to listen, boredom, and the desire for so much personal sovereignty over your time that you miss out on enriching experiences of communality (chapter 12). One useful approach for loosening your grip comes from the preschool education expert Tom Hobson, though, as he points out, its value is hardly limited to interactions with small children: when presented with a challenging or boring moment, try **deliberately adopting an attitude of curiosity**, in which your goal isn't to achieve any particular outcome, or successfully explain your position, but, as Hobson puts it, 'to figure out who this human being is that we're with'. Curiosity is a stance well suited to the inherent unpredictability of life with others, because it can be satisfied by their behaving in ways you like *or* dislike – whereas the stance of demanding a certain result is frustrated each time things fail to go your way.

Indeed, you could try taking this attitude towards everything, as the self-help writer Susan Jeffers suggests in her book *Embracing Uncertainty*. Not knowing what's coming next – which is the situation you're always in, with regard to the future – presents an ideal opportunity for choosing curiosity (*wondering* what might happen next) over worry (*hoping* that a certain specific thing will happen next, and fearing it might not) whenever you can.

9. Cultivate instantaneous generosity.

I'm definitely still working on the habit proposed (and practised) by the meditation teacher Joseph Goldstein: whenever a generous impulse arises in your mind – to give money, check in on a friend, send an email praising someone's work – **act on the impulse right away**, rather than putting it off until later. When we fail to act on such urges, it's rarely out of mean-spiritedness, or because we have second thoughts about whether the prospective recipient deserves it. More often, it's because of some attitude stemming from our efforts to feel in control of our time. We tell ourselves we'll turn to it when our urgent work is out of the way, or when we have enough spare time to do it really well; or that we ought first to spend a bit longer researching the best recipients for our charitable donations before making any, et cetera. But the only donations that count are the ones you actually get round to making. And while your colleague might appreciate a nicely worded message of praise more than a hastily worded one, the latter is vastly preferable to what's truly most likely to happen if you put it off, which is that you'll never get round to sending that message. All this takes some initial effort, but as Goldstein observes, the more selfish rewards are immediate, because generous action reliably makes you feel much happier.

10. Practise doing nothing.

'I have discovered that all the unhappiness of men arises from one single fact, that they cannot stay quietly in their own chamber,' Blaise Pascal wrote. When it comes to the challenge of using your four thousand weeks well, the capacity to do nothing is indispensable, because if you can't bear the discomfort of *not* acting, you're far more likely to make poor choices with your time, simply to feel as if you're acting – choices such as stressfully trying to hurry activities that won't be rushed (chapter 10) or feeling you ought to spend every moment being productive in the service of future goals, thereby postponing fulfilment to a time that never arrives (chapter 8).

Technically, it's impossible to do nothing at all: as long as you remain alive, you're always breathing, adopting some physical posture, and so forth. So training yourself to 'do nothing' really means training yourself to resist the urge to manipulate your experience or the people and things in the world around you – to let things be as they are. Shinzen Young teaches **'Do Nothing' meditation**, for which the instructions are to simply set a timer, probably only for five or ten minutes at first; sit down in a chair; and then stop trying to do anything. Every time you notice you're doing something – including thinking, or focusing on your breathing, or anything else – stop doing it. (If you notice you're criticising yourself inwardly for doing things, well, that's a thought, too, so stop doing that.) Keep on stopping until the timer goes off. 'Nothing is harder to do than nothing,' remarks the author and artist Jenny Odell. But to get better at it is to begin to regain your autonomy – to stop being motivated by the attempt to evade how reality feels here and now, to calm down, and to make better choices with your brief allotment of life.

Notes

Introduction: In the Long Run, We're All Dead

3 *the Frenchwoman who was thought to be 122 when she died in 1997:* Two decades after Jeanne Calment died, a pair of Russian researchers made the startling claim that 'Jeanne' was actually Yvonne, Jeanne's daughter, who had assumed her mother's identity upon her death years before. For the definitive account of the controversy – now largely settled in favour of the original version of events – see Lauren Collins, 'Living Proof', *New Yorker*, 17 and 24 February 2020.

4 *Biologists predict that lifespans within striking distance of Calment's:* For example, Bryan Hughes and Siegfried Hekimi, 'Many Possible Maximum Lifestyle Trajectories', *Nature* 546 (2017): E8–E9.

4 *'This space that has been granted to us':* Seneca, 'De Brevitate Vitae', in *Moral Essays*, vol. 2, trans. John W. Basore (Cambridge, MA: Loeb Classical Library, 1932), 287.

4 *'we will all be dead any minute':* Thomas Nagel, 'The Absurd', *Journal of Philosophy* 68 (1971): 716–27.

6 *Surveys reliably show that we feel more pressed for time:* See Jonathan Gershuny, 'Busyness as the Badge of Honor for the New Superordinate Working Class', *Social Research* 72 (2005): 287–315.

6 *in 2013, research by a team of Dutch academics:* Anina Vercruyssen et al., 'The Effect of Busyness on Survey Participation: Being Too Busy or Feeling Too Busy to Cooperate?', *International Journal of Social Research Methodology* 17 (2014): 357–71.

7 *It's because our time and attention are so limited:* See James Williams, *Stand Out of Our Light: Freedom and Resistance in the Attention Economy* (Cambridge: Cambridge University Press, 2018).

8 *'in a new kind of everlasting present':* Fredrick Matzner, quoted in Matt Simon, 'Why Life During a Pandemic Feels So Surreal', *Wired*, 31 March

2020, available at www.wired.com/story/why-life-during-a-pandemic-feels-so-surreal/.

9 *the American anthropologist Edward T. Hall once pointed out:* Edward T. Hall, *The Dance of Life: The Other Dimension of Time* (New York: Anchor, 1983), 84.

9 *'a generation of finely honed tools':* Malcolm Harris, *Kids These Days: The Making of Millennials* (New York: Back Bay Books, 2018), 76.

10 *it is 'possible for a person to have an overwhelming number of things to do':* David Allen, *Getting Things Done: The Art of Stress-Free Productivity* (New York: Penguin, 2015), 3.

10 *'what the martial artists call a "mind like water"':* Allen, *Getting Things Done*, 11.

11 *'For the first time since his creation':* John Maynard Keynes, 'Economic Possibilities for Our Grandchildren' (1930), downloaded from www.econ.yale.edu/smith/econ116a/keynes1.pdf.

12 *'Life, I knew, was supposed to be more joyful than this':* Charles Eisenstein, *The More Beautiful World Our Hearts Know Is Possible* (Berkeley, CA: North Atlantic Books, 2013), 2.

13 *'The spirit of the times is one of joyless urgency':* Marilynne Robinson, *The Givenness of Things: Essays* (New York: Farrar, Straus and Giroux, 2015), 4.

1. The Limit-Embracing Life

18 *St Anthony's fire:* See Ángel Sánchez-Crespo, 'Killer in the Rye: St. Anthony's Fire', *National Geographic*, 27 November 2018, available at www.nationalgeographic.com/history/magazine/2018/11–12/ergotism-infections-medieval-europe/.

19 *'an independent world of mathematically measurable sequences':* Lewis Mumford, *Technics and Civilization* (Chicago: University of Chicago Press, 2010), 15.

21 *Medieval people might speak of a task lasting a 'Miserere whyle':* E. P. Thompson, 'Time, Work-Discipline, and Industrial Capitalism,' *Past and Present* 38 (1967): 81.

21 *Richard Rohr, a contemporary Franciscan priest and author:* Richard Rohr, 'Living in Deep Time', *On Being* podcast, available at https://www.wnyc.org/story/richard-rohr--living-in-deep-time/.

21 *'into a realm where there is enough of everything':* Gary Eberle, *Sacred Time and the Search for Meaning* (Boston: Shambhala, 2002), 7.

21 *'The clock does not stop, of course':* Eberle, *Sacred Time and the Search for Meaning*, 8.

22 *'From a low hill in this broad savanna':* Carl Jung, *Memories, Dreams, Reflections* (New York: Vintage, 1989), 255.

24 *'I have by sundry people [been] horribly cheated':* Thompson, 'Time, Work-Discipline, and Industrial Capitalism', 81. I have modernised the spelling here.

25 *'One thinks with a watch in one's hand':* Friedrich Nietzsche, *The Gay Science* (New York: Vintage, 1974), 259.

25 *a book that arrived on my desk the other day:* This is Brian Tracy, *Master Your Time, Master Your Life: The Breakthrough System to Get More Results, Faster, in Every Area of Your Life* (New York: TarcherPerigee, 2016).

26 *'Eternity ceased gradually to serve as the measure and focus':* Mumford, *Technics and Civilization*, 14.

30 *'we don't have to consciously participate':* Bruce Tift, *Already Free: Buddhism Meets Psychotherapy on the Path of Liberation* (Boulder: Sounds True, 2015), 152.

30 *'We labour at our daily work more ardently':* Friedrich Nietzsche, *Untimely Meditations* (Cambridge: Cambridge University Press, 1997), 158.

33 *'You teach best what you most need to learn':* Richard Bach, *Illusions: The Adventures of a Reluctant Messiah* (New York: Delta, 1998), 48.

33 *what in German has been called* Eigenzeit*:* Morten Svenstrup, *Towards a New Time Culture*, trans. Peter Holm-Jensen (Copenhagen: Author, 2013), 8.

34 *as the journalist Anne Helen Petersen writes:* Anne Helen Petersen, 'How Millennials Became the Burnout Generation', BuzzFeed, 5 January 2019, available at www.buzzfeednews.com/article/annehelenpetersen/millennials-burnout-generation-debt-work.

35 *'Depressing? Not a bit of it':* Charles Garfield Lott Du Cann, *Teach Yourself to Live* (London: Teach Yourself, 2017), loc. 107 of 2101, Kindle.

2. The Efficiency Trap

38 *Research shows that this feeling arises on every rung of the economic ladder:* On the ways 'time poverty' and economic poverty interact, see for example Andrew S. Harvey and Arun K. Mukhopadhyay, 'When Twenty-Four Hours Is Not Enough: Time Poverty of Working Parents', *Social Indicators Research* 82 (2007): 57–77. But feelings of (and complaints about) busyness are actually worse among those earning more: See Daniel Hammermesh, *Spending Time: The Most Valuable Resource* (New York: Oxford University Press, 2018).

38 *As the law professor Daniel Markovits has shown:* Daniel Markovits, 'How Life Became an Endless, Terrible Competition', *The Atlantic*, September 2019, available at www.theatlantic.com/magazine/archive/2019/09/meritocracys-miserable-winners/594760/.

39 *the English journalist Arnold Bennett published a short and grouchy book:* All quotations from *How to Live on 24 Hours a Day* are from the unpaginated Project Gutenberg transcription, available at www.gutenberg.org/files/2274/2274-h/2274-h.htm.

42 *in her book* More Work for Mother*:* Ruth Schwartz Cowan, 'The Invention of Housework: The Early Stages of Industrialization', in *More Work for Mother: The Ironies of Household Technology from the Open Hearth to the Microwave* (London: Free Association, 1989), 40–68.

42 *'Work expands so as to fill the time available':* C. Northcote Parkinson, 'Parkinson's Law,' *The Economist*, 19 November 1955, available at www.economist.com/news/1955/11/19/parkinsons-law.

45 *As the German sociologist Hartmut Rosa explains:* Hartmut Rosa, *Social Acceleration: A New Theory of Modernity*, trans. Jonathan Trejo-Mathys (New York: Columbia University Press, 2015).
46 *'The more we can accelerate our ability to go to different places':* Jonathan Trejo-Mathys, 'Translator's Introduction', in Rosa, *Social Acceleration*, xxi.
49 *'a limitless reservoir for other people's expectations':* Jim Benson, personal communication.
53 *'don't even realize something is broken':* Alexis Ohanian, *Without Their Permission: How the 21st Century Will Be Made, Not Managed* (New York: Business Plus, 2013), 159.
53 *'I prefer to brew my coffee':* Tim Wu, 'The Tyranny of Convenience', *New York Times*, 18 February 2018.
54 *'Every morning I carefully scrape out the ash of yesterday':* Sylvia Keesmaat, 'Musings on an Inefficient Life', *Topology*, 16 March 2017, available at www.topologymagazine.org/essay/throwback/musings-on-an-inefficient-life/.
55 *'How else are we to get to know this place':* Keesmaat, 'Musings on an Inefficient Life'.

3. Facing Finitude

57 *'Being-towards-death':* Martin Heidegger, *Being and Time*, trans. John Macquarrie and Edward Robinson (Oxford: Blackwell, 1962), 277 and passim.
57 *'de-severance':* Heidegger, *Being and Time*, 139.
57 *'anxiety "in the face of" that potentiality-for-Being':* Heidegger, *Being and Time*, 295.
58 *'a world is worlding all around us':* Martin Heidegger, quoted in Richard Polt, *Heidegger: An Introduction* (Ithaca, NY: Cornell University Press, 1999), 1.
58 *'the brute reality on which all of us ought to be constantly stubbing our toes':* Sarah Bakewell, *At the Existentialist Café: Freedom, Being, and Apricot Cocktails* (New York: Other Press, 2016), 51.
62 *'If I believed that my life would last forever':* Martin Hägglund, *This Life: Why Mortality Makes Us Free* (London: Profile, 2019), 5.
62 *'Heaven: Will It Be Boring?':* Quoted in Hägglund, *This Life*, 4.
64 *'Something has happened. A piece of news':* Marion Coutts, *The Iceberg: A Memoir* (New York: Black Cat, 2014), loc. 23 of 3796, Kindle.
65 *'bright sadness':* Richard Rohr, *Falling Upward: A Spirituality for the Two Halves of Life* (San Francisco: Jossey-Bass, 2011), 117.
65 *'stubborn gladness':* A paraphrasing of Jack Gilbert's poem 'A Brief for the Defense', published in *Collected Poems* (New York: Knopf, 2014), 213.
65 *'sober joy':* Bruce Ballard, review of '*Heidegger's Moral Ontology* by James Reid', *Review of Metaphysics* 73 (2020): 625–26.
65 *'a sort of personal affront':* Paul Sagar, 'On Going On and On and On,' *Aeon*, 3 September 2018, available at aeon.co/essays/theres-a-big-problem-with-immortality-it-goes-on-and-on.

66 *'I was early . . . so I spent some time in a nearby park':* All quotations from David Cain in this chapter come from 'Your Whole Life Is Borrowed Time', *Raptitude*, 13 August 2018, available at www.raptitude.com/2018/08/your-whole-life-is-borrowed-time.

4. Becoming a Better Procrastinator

71 *As the American author and teacher Gregg Krech puts it:* Gregg Krech, *The Art of Taking Action: Lessons from Japanese Psychology* (Monkton, VT: ToDo Institute, 2014), 19.

72 *the extraordinarily irritating parable of the rocks in the jar:* Stephen R. Covey, *First Things First* (New York: Free Press, 1996), 88.

73 *the graphic novelist and creativity coach Jessica Abel:* Quotations from Jessica Abel come from 'How to Escape Panic Mode and Embrace Your Life-Expanding Projects', available at jessicaabel.com/pay-yourself-first-life-expanding-projects/.

75 *In their book* Personal Kanban*:* Jim Benson and Tonianne DeMaria Barry, *Personal Kanban: Mapping Work, Navigating Life* (Scotts Valley, CA: CreateSpace, 2011), 39.

77 *There is a story attributed to Warren Buffett:* The tale of the purported origins of this story, and of Buffett's comment that he can't recall anything of the sort, is related in Ruth Umoh, 'The Surprising Lesson This 25-Year-Old Learned from Asking Warren Buffett an Embarrassing Question', CNBC Make It, 5 June 2018, available at www.cnbc.com/2018/06/05/warren-buffetts-answer-to-this-question-taught-alex-banayan-a-lesson.html.

78 *'it's much harder than that':* Elizabeth Gilbert attributes this line to 'a wise older woman' in a Facebook post dated 4 November 2015, available at www.facebook.com/GilbertLiz/posts/how-many-times-in-your-life-have-you-needed-to-say-thisand-do-you-need-to-say-it/915704835178299/.

79 *The philosopher Costica Bradatan illustrates the point:* Costica Bradatan, 'Why Do Anything? A Meditation on Procrastination', *New York Times*, 18 September 2016.

80 *consider the case of the worst boyfriend ever, Franz Kafka:* In addition to the original letters, reproduced in *Letters to Felice*, ed. Erich Heller and Jürgen Born (New York: Schocken, 1973), my account of Kafka's relationship with Felice Bauer draws on Eleanor Bass, 'Kafka Was a Terrible Boyfriend', LitHub, 14 February 2018, available at lithub.com/kafka-was-a-terrible-boyfriend; and Rafia Zakaria, 'Franz Kafka's Virtual Romance: A Love Affair by Letters as Unreal as Online Dating', *Guardian* books blog, 12 August 2016, available at www.theguardian.com/books/booksblog/2016/aug/12/franz-kafkas-virtual-world-romance-felice-bauer.

81 *'neuroses are no different from ours':* Morris Dickstein, 'A Record of Kafka's Love for a Girl and Hate for Himself', *New York Times*, 30 September 1973.

83 *'the future, which we dispose of to our liking':* Henri Bergson, *Time and Free Will: An Essay on the Immediate Data of Consciousness*, trans. F. L. Pogson (Mineola, NY: Dover, 2001), 9.

83 *'The idea of the future, pregnant with an infinity of possibilities':* Bergson, *Time and Free Will*, 10.

85 *'You must settle, in a relatively enduring way':* Robert E. Goodin, *On Settling* (Princeton, NJ: Princeton University Press, 2012), 65.

87 *Once, in an experiment, the Harvard University social psychologist Daniel Gilbert:* Daniel Gilbert and Jane Ebert, 'Decisions and Revisions: The Affective Forecasting of Changeable Outcomes', *Journal of Personality and Social Psychology* 82 (2002): 503–14.

5. The Watermelon Problem

89 *watching two reporters from BuzzFeed wrap rubber bands around a watermelon:* Chelsea Marshall, James Harness and Edd Souaid, 'This Is What Happens When Two BuzzFeed Employees Explode a Watermelon', BuzzFeed, 8 April 2016, available at www.buzzfeed.com/chelseamarshall/watermelon-explosion.

90 *'I want to stop watching so bad':* 'In Online First, "Exploding Watermelon" Takes the Cake', Phys.org, 8 April 2016, available at phys.org/news/2016–04-online-watermelon-cake.html.

90 *'I've been watching you guys put rubber bands around a watermelon':* Tasneem Nashrulla, 'We Blew Up a Watermelon and Everyone Lost Their Freaking Minds', BuzzFeed, 8 April 2016, available at www.buzzfeednews.com/article/tasneemnashrulla/we-blew-up-a-watermelon-and-everyone-lost-their-freaking-min.

91 *according to one calculation, by the psychologist Timothy Wilson:* Quoted in Jane Porter, 'You're More Biased Than You Think', *Fast Company*, 6 October 2014, available at www.fastcompany.com/3036627/youre-more-biased-than-you-think.

92 *'baking their bodies in the sun':* Seneca, 'De Brevitate Vitae', in *Moral Essays*, vol. 2, trans. John W. Basore (Cambridge, MA: Loeb Classical Library, 1932), 327.

93 *the case of the Austrian psychotherapist Viktor Frankl:* Viktor Frankl, *Man's Search for Meaning* (Boston: Beacon, 2006).

94 *'Attention is the beginning of devotion':* Mary Oliver, *Upstream: Selected Essays* (New York: Penguin, 2016), loc. 166 of 1669, Kindle.

95 *the former Facebook investor turned detractor Roger McNamee:* Quoted in 'Full Q&A: *Zucked* Author Roger McNamee on *Recode Decode*', *Vox*, 11 February 2019, available at www.vox.com/podcasts/2019/2/11/18220779/zucked-book-roger-mcnamee-decode-kara-swisher-podcast-mark-zuckerberg-facebook-fb-sheryl-sandberg.

96 *In the words of the philosopher:* Quoted in James Williams, *Stand Out of Our Light* (Cambridge: Cambridge University Press, 2018), xii.

98 *'distracted from distraction by distraction':* T. S. Eliot, 'Burnt Norton', in *Four Quartets* (Boston: Mariner, 1968), 5.

99 *'a thousand people on the other side of the screen':* For example, in Bianca Bosker, 'The Binge Breaker', *The Atlantic*, November 2016, available at www.theatlantic.com/magazine/archive/2016/11/the-binge-breaker/501122/.

6. The Intimate Interrupter

101 *during the winter months of 1969:* My account of Steve/Shinzen Young's story, and all quotations from Young, come from my interview with him and from Shinzen Young, *The Science of Enlightenment: How Meditation Works* (Boulder: Sounds True, 2016).

104 *'the intimate interrupter':* Oliver, *Upstream: Selected Essays*, loc. 305 of 1669, Kindle.

104 *'self within the self':* Oliver, *Upstream: Selected Essays*, loc. 302 of 1669, Kindle.

104 *'One of the puzzling lessons I have learned':* Krech, *The Art of Taking Action*, 71.

106 *To quote the psychotherapist Bruce Tift once more:* Tift, *Already Free*, 152.

106 *'a realm in which space doesn't matter and time spreads out':* James Duesterberg, 'Killing Time', *The Point Magazine*, 29 March 2020, available at thepointmag.com/politics/killing-time/.

109 *Some Zen Buddhists hold:* See, for example, John Tarrant, 'You Don't Have to Know', *Lion's Roar*, 7 March 2013, available at www.lionsroar.com/you-dont-have-to-know-tales-of-trauma-and-transformation-march-2013/.

7. We Never Really Have Time

113 *'Hofstadter's law':* Douglas Hofstadter, *Gödel, Escher, Bach: An Eternal Golden Braid* (New York: Basic Books, 1999), 152.

115 *'Dad Suggests Arriving at Airport 14 Hours Early':* *The Onion*, 22 September 2012, available at www.theonion.com/dad-suggests-arriving-at-airport-14-hours-early-1819573933.

117 *'We assume we have three hours or three days to do something':* David Cain, 'You Never Have Time, Only Intentions', *Raptitude*, 23 May 2017, available at www.raptitude.com/2017/05/you-never-have-time-only-intentions.

119 *'So imprudent are we':* Blaise Pascal, *Pensées*, trans. W. F. Trotter (Mineola, NY: Dover, 2018), 49.

120 *'If I go to sleep after lunch in the room where I work':* Simone de Beauvoir, *All Said and Done*, trans. Patrick O'Brian (New York: Putnam, 1974), 1.

121 *'Trying to control the future is like trying to take the master carpenter's place':* Stephen Mitchell, *Tao Te Ching: A New English Version* (New York: Harper Perennial Modern Classics, 2006), 92.

121 *'Do not rule over imaginary kingdoms of endlessly proliferating possibilities':* Quoted in Shaila Catherine, 'Planning and the Busy Mind', talk transcript available at www.imsb.org/teachings/written-teachings-articles-and-interviews/planning-and-the-busy-mind-2.

121 *'Take no thought for the morrow':* Matthew 6:34, the Bible: King James Version (London: Penguin Classics, 2006), 1555.

122 *'Partway through this particular talk . . . Krishnamurti suddenly paused':* Quoted in Bhava Ram, *Deep Yoga: Ancient Wisdom for Modern Times* (Coronado, CA: Deep Yoga, 2013), 76.

123 *in the words of the American meditation teacher Joseph Goldstein:* Quoted in Catherine, 'Planning and the Busy Mind'.

8. You Are Here

126 *In his book* Back to Sanity, *the psychologist Steve Taylor:* Steve Taylor, *Back to Sanity* (London: Hay House, 2012), 61.

126 *what I once heard described as the '"when-I-finally" mind':* Tara Brach, personal communication.

127 *'Take education. What a hoax':* Alan Watts, 'From Time to Eternity', in *Eastern Wisdom, Modern Life: Collected Talks 1960–1969* (Novato, CA: New World Library, 2006), 109–10.

129 *there's one West African ethnic group, the Hausa-Fulani:* Robert A. LeVine and Sarah LeVine, *Do Parents Matter? Why Japanese Babies Sleep Soundly, Mexican Siblings Don't Fight, and American Families Should Just Relax* (New York: PublicAffairs, 2016), x.

131 *The writer Adam Gopnik calls the trap into which I had fallen the 'causal catastrophe':* Adam Gopnik, 'The Parenting Paradox', *New Yorker*, 29 January 2018.

132 *'Because children grow up, we think a child's purpose is to grow up':* Tom Stoppard, *The Coast of Utopia* (New York: Grove Press, 2007), 223.

132 *But the author and podcast host Sam Harris makes the disturbing observation:* Sam Harris, 'The Last Time', a talk in the Waking Up app, available at www.wakingup.com.

134 *Mexico, for example, has often outranked the United States in global indices:* See, for example, the Happy Planet Index, at happyplanetindex.org; and John Helliwell, Richard Layard and Jeffrey Sachs, eds, *World Happiness Report 2013* (New York: UN Sustainable Development Solutions Network, 2013).

135 *'Lawyers imbued with the ethos of the billable hour':* M. Cathleen Kaveny, 'Billable Hours and Ordinary Time: A Theological Critique of the Instrumentalization of Time in Professional Life', *Loyola University of Chicago Law Journal* 33 (2001): 173–220.

136 *'The "purposive" man . . . is always trying to secure a spurious and delusive immortality':* John Maynard Keynes, 'Economic Possibilities for Our Grandchildren' (1930), downloaded from www.econ.yale.edu/smith/econ116a/keynes1.pdf.

137 *'[We] see the Crater Lake with a feeling of, "Well, there it is,"':* Robert M. Pirsig, *Zen and the Art of Motorcycle Maintenance* (New York: William Morrow, 1974), 341.

139 *that quotation from the bestselling Buddhist teacher Thich Nhat Hanh:* Thich Nhat Hanh, *The Miracle of Mindfulness*, trans. Mobi Ho (Boston: Beacon, 1999), 3.

139 *the 2015 study by researchers at Carnegie Mellon University in Pittsburgh:* George Loewenstein et al., 'Does Increased Sexual Frequency Enhance Happiness?', *Journal of Economic Behavior and Organization* (2015): 206–18.

140 *'We cannot get anything out of life'*: Jay Jennifer Matthews, *Radically Condensed Instructions for Being Just as You Are* (Scotts Valley, CA: CreateSpace, 2011), 27, emphasis added.

9. Rediscovering Rest

143 *'Relax! You'll Be More Productive':* Tony Schwartz, 'Relax! You'll Be More Productive', *New York Times*, 10 February 2013.

143 *'We are all of us compelled . . . to read for profit':* Walter Kerr, quoted in Staffan Linder, *The Harried Leisure Class* (New York: Columbia University Press, 1970), 4.

143 *we actually have more leisure time than we did in previous decades:* See, for example, J. H. Ausuble and A. Gruebler, 'Working Less and Living Longer: Long-Term Trends in Working Time and Time Budgets', *Technological Forecasting and Social Change* 50 (1995): 113–31.

144 *research suggests that this problem grows worse the wealthier you get:* Daniel Hamermesh's research is discussed in Allana Akhtar, 'Wealthy Americans Don't Have Enough Time in the Day to Spend Their Money, and It's Stressing Them Out', *Business Insider*, 26 June 2019, available at markets.businessinsider.com/news/stocks/how-the-desire-for-status-symbols-leads-to-stress-2019-6-1028309783.

145 *Some historians claim that the average country-dweller:* Juliet Shor, *The Overworked American* (New York: Basic Books, 1992), 47.

145 *'The laboring man . . . will take his rest long in the morning':* Quoted in Shor, *The Overworked American*, 43.

147 *To 'look around to see what is going on':* Livia Gershon, 'Clocking Out', *Longreads*, July 2018, available at longreads.com/2018/07/11/clocking-out/.

147 The Right To Be Lazy*:* Paul Lafargue, *The Right To Be Lazy* (1883), available at www.marxists.org/archive/lafargue/1883/lazy/.

147 *'If the satisfaction of an old man drinking a glass of wine counts for nothing':* Simone de Beauvoir, *The Ethics of Ambiguity* (New York: Open Road, 2015), 146.

148 *'I don't get to bed until I'm so tired I could sleep on the floor':* All quotes from Danielle Steel come from Samantha Leach, 'How the Hell Has Danielle Steel Managed to Write 179 Books?', *Glamour*, 9 May 2019, available at www.glamour.com/story/danielle-steel-books-interview.

149 *Social psychologists call this inability to rest 'idleness aversion':* C. K. Hsee et al., 'Idleness Aversion and the Need for Justifiable Busyness', *Psychological Science* 21 (2010): 926–30.

149 *his famous theory of the 'Protestant work ethic':* Max Weber, *The Protestant Ethic and the Spirit of Capitalism and Other Writings* (London: Penguin Classics, 2002).

150 *guilty sinners anxious to expunge the stain of laziness:* I owe this thought to David Zahl, *Seculosity: How Career, Parenting, Technology, Food, Politics, and Romance Became Our New Religion and What to Do About It* (Minneapolis: Fortress Press, 2019), 106–7.

151 *'We are the sum of all the moments of our lives':* Thomas Wolfe, *Look Homeward, Angel* (New York: Simon & Schuster, 1995), xv.

152 *'Most people mistakenly believe that all you have to do to stop working is not work':* Judith Shulevitz, 'Bring Back the Sabbath', *New York Times*, 2 March 2003.

153 *In his book* Sabbath as Resistance*:* Walter Brueggemann, *Sabbath as Resistance: Saying No to the Culture of Now* (Louisville, KY: Westminster John Knox Press, 2014), xiv.

154 *'Nothing is more alien to the present age than idleness':* John Gray, *Straw Dogs: Thoughts on Humans and Other Animals* (New York: Farrar, Straus and Giroux, 2002), 195.

154 *'How can there be play':* Gray, *Straw Dogs*, 196.

156 *'You can stop doing these things, and you eventually will':* Kieran Setiya, *Midlife: A Philosophical Guide* (Princeton, NJ: Princeton University Press, 2017), 134.

157 *'If, on the other hand, [the human animal] lacks objects of willing':* Quoted in Setiya, *Midlife*, 131.

159 *My respect for the rock star Rod Stewart:* Steve Flint and Craig Tiley, 'In My Heart, and in My Soul: Sir Rod Stewart on His Lifelong Love of Model Railways', *Railway Modeller*, December 2019.

160 *The publisher and editor Karen Rinaldi:* Karen Rinaldi, '(It's Great to) Suck at Something', *New York Times*, 28 April 2017.

10. The Impatience Spiral

162 *The practice of inching towards the car in front:* S. Farzad Ahmadi et al., 'Latent Heat of Traffic Moving from Rest', *New Journal of Physics* 19 (2017), available at iopscience.iop.org/article/10.1088/1367–2630/aa95f0.

163 *It has been calculated that if Amazon's front page loaded one second more slowly:* See Kit Eaton, 'How One Second Could Cost Amazon $1.6 Billion in Sales', *Fast Company*, 15 March 2012, available at www.fastcompany.com/1825005/how-one-second-could-cost-amazon-16-billion-sales.

165 *'I've been finding it harder and harder to concentrate on words':* Hugh McGuire, 'Why Can't We Read Anymore?', *Medium*, 22 April 2015, available at medium.com/@hughmcguire/why-can-t-we-read-anymore-503c38c131fe.

165 *'It is not simply that one is interrupted':* Tim Parks, 'Reading: The Struggle', *New York Review of Books*, *NYR Daily* blog, 10 June 2014, available at www.nybooks.com/daily/2014/06/10/reading-struggle/.

166 *a psychotherapist in California named Stephanie Brown:* All quotations from Stephanie Brown come from my interview with Brown and from Stephanie Brown, *Speed: Facing Our Addiction to Fast and Faster – and Overcoming Our Fear of Slowing Down* (New York: Berkley, 2014).

169 *As the science writer James Gleick points out:* James Gleick, *Faster: The Acceleration of Just About Everything* (New York: Pantheon, 1999), 12.

169 *'We admitted,' reads the first of the Twelve Steps:* The twelve steps of Alcoholics Anonymous are available at www.alcohol.org/alcoholics-anonymous.

11. Staying on the Bus

174 *I first learned this lesson from Jennifer Roberts:* All quotations from Jennifer Roberts come from my interview with Roberts and from Jennifer Roberts, 'The Power of Patience', *Harvard Magazine*, November–December 2013, available at https://harvardmagazine.com/2013/11/the-power-of-patience.

177 *'tangible, almost edible':* Robert Grudin, *Time and the Art of Living* (Cambridge: Harper and Row, 1982), 125.

178 *'Boy, I sure admire you':* All quotations from M. Scott Peck come from 'Problem-Solving and Time', in *The Road Less Travelled: A New Psychology of Love, Traditional Values and Spiritual Growth* (London: Arrow Books, 2006), 15–20.

182 *'includes a big component of impatience about not being finished':* Robert Boice, *How Writers Journey to Comfort and Fluency: A Psychological Adventure* (Westport, CT: Praeger, 1994), 33.

182 *The Finnish American photographer Arno Minkkinen dramatises this deep truth:* A transcript of Minkkinen's 2004 commencement address, 'Finding Your Own Vision', at the New England School of Photography, where he outlines this theory, is available at jamesclear.com/great-speeches/finding-your-own-vision-by-arno-rafael-minkkinen.

12. The Loneliness of the Digital Nomad

186 *'I don't have to take out the garbage':* All quotations from Mario Salcedo come from Lance Oppenheim, 'The Happiest Guy in the World', *New York Times*, 1 May 2018, available at www.nytimes.com/2018/05/01/opinion/cruise-caribbean-retirement.html.

188 *'A person with a flexible schedule and average resources':* Scott Adams, *How to Fail at Almost Everything and Still Win Big: Kind of the Story of My Life* (New York: Portfolio, 2013), 173.

189 *'Last year, I visited 17 countries':* Mark Manson, 'The Dark Side of the Digital Nomad', available at markmanson.net/digital-nomad.

189 *In 2013, a researcher from Uppsala in Sweden named:* Terry Hartig et al., 'Vacation, Collective Restoration, and Mental Health in a Population', *Society and Mental Health* 3 (2013): 221–36.

191 *research, which has demonstrated that people in long-term unemployment get a happiness boost:* Cristobal Young and Chaeyoon Lim, 'Time as a Network Good: Evidence from Unemployment and the Standard Workweek', *Sociological Science* 1 (2014): 10–27.

192 *The historian Clive Foss has described the nightmare that transpired:* Clive Foss, 'Stalin's Topsy-Turvy Work Week', *History Today*, September 2004. I have also drawn here on Judith Shulevitz, 'Why You Never See Your Friends Anymore', *The Atlantic*, November 2019.

193 *'Lenin's widow, in good Marxist fashion':* E. G. Richards, *Mapping Time: The Calendar and Its History* (Oxford: Oxford University Press, 2000), 278.

193 *'What are we to do at home if the wife is in the factory':* Quoted in Shulevitz, 'Why You Never See Your Friends Anymore'.

195 *'Marching aimlessly about on the drill field':* William H. McNeill, *Keeping Together in Time: Dance and Drill in Human History* (Cambridge, MA: Harvard University Press, 1995), 2.

196 *And some evolutionary biologists speculate:* See Jay Schulkin and Greta Raglan, 'The Evolution of Music and Human Social Capability', *Frontiers in Neuroscience* 8 (2014): 292.

196 *a study based on a frame-by-frame analysis of the race:* Manuel Varlet and Michael J. Richardson, 'What Would Be Usain Bolt's 100-Meter Sprint World Record Without Tyson Gay? Unintentional Interpersonal Synchronization Between the Two Sprinters', *Journal of Experimental Psychology: Human Perception and Performance* 41 (2015): 36–41.

197 *The extraordinary psychological benefits of choral singing:* Betty Bailey and Jane Davidson, 'Effects of Group Singing and Performance for Marginalized and Middle-Class Singers', *Psychology of Music* 33 (2005): 269–303.

197 *The world 'doesn't open up into a million shimmering dimensions':* Stacy Horn, 'Ode to Joy', *Slate*, 25 July 2013, available at slate.com/human-interest/2013/07/singing-in-a-choir-research-shows-it-increases-happiness.html.

200 *'Totalitarian movements are mass organizations of atomized, isolated individuals':* Hannah Arendt, *The Origins of Totalitarianism* (New York: Harvest, 1973), 323.

13. Cosmic Insignificance Therapy

203 *The Jungian psychotherapist James Hollis recalls:* In James Hollis, *Finding Meaning in the Second Half of Life: How to Finally, Really Grow Up* (New York: Gotham, 2005), 2.

204 *'Then I considered all that my hands had done':* Ecclesiastes 2:11, the Bible: English Standard Version (Wheaton, IL: Crossway, 2005), 471.

206 *'What the trauma has shown us':* All quotations from Julio Vincent Gambuto are from 'Prepare for the Ultimate Gaslighting', *Medium,* 10 April 2020, available at forge.medium.com/prepare-for-the-ultimate-gaslighting-6a8ce3f0a0e0.

208 *The late British philosopher Bryan Magee liked to make the following arresting point:* Bryan Magee, *Ultimate Questions* (Princeton, NJ: Princeton University Press, 2016), 1–2.

209 *'the number of friends I squeeze into my living room':* Magee, *Ultimate Questions*, 2.

210 *'the massive indifference of the universe':* Richard Holloway, *Looking in the Distance* (Edinburgh: Canongate, 2005), 13.

210 The Universe Doesn't Give a Flying Fuck About You*:* Johnny Truant, *The Universe Doesn't Give a Flying Fuck About You.* Self-published, Amazon Digital Services, 2014. Kindle.

211 *'transcend the common and the mundane':* Iddo Landau, *Finding Meaning in an Imperfect World* (New York: Oxford University Press, 2017), 31.

212 *'We do not disapprove of a chair because it cannot be used to boil water':* Landau, *Finding Meaning in an Imperfect World*, 39.

212 *'implausible, for almost all people, to demand of themselves':* Landau, *Finding Meaning in an Imperfect World*, 39.

14. The Human Disease

217 *'Time is the substance I am made of':* Jorge Luis Borges, 'A New Refutation of Time', in *Labyrinths* (New York: New Directions, 2007), 234.

218 *'There is a strange attitude and feeling that one is* not yet *in real life':* Marie-Louise von Franz, *The Problem of the Puer Aeternus* (Toronto: Inner City), 8.

219 *as the Zen teacher Charlotte Joko Beck puts it:* Quoted in Joan Tollifson, *Death: The End of Self-Improvement* (Salisbury, UK: New Sarum Press, 2019), 60.

219 *'I was peeling a red apple from the garden':* Christian Bobin quoted in Christophe André, *Looking at Mindfulness: Twenty-Five Paintings to Change the Way You Live* (New York: Blue Rider, 2011), 256.

220 *'live the questions':* Rainer Maria Rilke, *Letters to a Young Poet* (New York: W. W. Norton, 2004), 27.

221 *James Hollis recommends asking:* James Hollis, *What Matters Most: Living a More Considered Life* (New York: Gotham, 2009), 13.

222 *There is a sort of cruelty:* Landau, *Finding Meaning in an Imperfect World*, 40–41.

223 *'at a certain age . . . it finally dawns on us that, shockingly, no one really* cares*':* Stephen Cope, *The Great Work of Your Life: A Guide for the Journey to Your True Calling* (New York: Bantam, 2015), 37.

224 *The Buddhist teacher Susan Piver points out:* Susan Piver, 'Getting Stuff Done by Not Being Mean to Yourself', 20 August 2010, available at openheartproject.com/getting-stuff-done-by-not-being-mean-to-yourself.

226 *in his documentary* A Life's Work*:* David Licata, *A Life's Work* (2019), at alifesworkmovie.com.

227 *'Dear Frau V. . . . Your questions are unanswerable':* Carl Jung, *Letters*, vol. 1, *1906–1950* (Oxford: Routledge, 2015), 132.

Afterword: Beyond Hope

230 *People sometimes ask Derrick Jensen:* All quotations from Derrick Jensen come from 'Beyond Hope', *Orion*, https://orionmagazine.org/article/beyond-hope/.

231 *'there's always going to be a babysitter available when we need one':* Pema Chödrön, *When Things Fall Apart* (Boulder: Shambhala, 2016), 38.
232 *'People say, "Oh, when the apocalypse comes . . ."':* Nellie Bowles, 'Fleeing Babylon for a Wild Life', *New York Times*, 5 March 2020.
233 *'Abandoning hope is an affirmation':* Chödrön, *When Things Fall Apart*, 40.
234 *'Spring is here, even in London N1':* George Orwell, 'Some Thoughts on the Common Toad', first published in *Tribune*, 12 April 1946, available at www.orwellfoundation.com/the-orwell-foundation/orwell/essays-and-other-works/some-thoughts-on-the-common-toad/.

Appendix: Ten Tools for Embracing Your Finitude

236 *'You could fill any arbitrary number of hours with what feels to be productive work':* Cal Newport, 'Fixed-Schedule Productivity: How I Accomplish a Large Amount of Work in a Small Number of Work Hours', available at www.calnewport.com/blog/2008/02/15/fixed-schedule-productivity-how-i-accomplish-a-large-amount-of-work-in-a-small-number-of-work-hours/, with further discussion in Cal Newport, *Deep Work* (New York: Grand Central, 2016).
238 *'When you can't do it all, you feel ashamed and give up':* Jon Acuff, *Finish: Give Yourself the Gift of Done* (New York: Portfolio, 2017), 36.
239 *evidence for the motivating power of 'small wins':* See Teresa Amabile and Steven Kramer, *The Progress Principle: Using Small Wins to Ignite Joy, Engagement, and Creativity at Work* (Brighton, MA: Harvard Business Review Press, 2011).
240 *'After going to grayscale, I'm not a different person':* Nellie Bowles, 'Is the Answer to Phone Addiction a Worse Phone?', *New York Times*, 12 January 2018.
241 *'As each passing year converts . . . experience into automatic routine':* William James, *The Principles of Psychology*, vol. 1 (New York: Dover, 1950), 625.
242 *'your experience of life would be* twice as full *as it currently is':* Young, *The Science of Enlightenment*, 31.
243 *'to figure out who this human being is that we're with':* Tom Hobson in conversation with Janet Lansbury, 'Stop Worrying About Your Preschooler's Education', available at www.janetlansbury.com/2020/05/stop-worrying-about-your-preschoolers-education.
243 *Susan Jeffers suggests in her book* Embracing Uncertainty*:* Susan Jeffers, *Embracing Uncertainty: Breakthrough Methods for Achieving Peace of Mind When Facing the Unknown* (New York: St. Martin's Press, 2003).
244 *'I have discovered that all the unhappiness of men':* Pascal, *Pensées*, 49.
245 *'Nothing is harder to do than nothing':* Jenny Odell, *How to Do Nothing* (New York: Melville House, 2019), ix.

Acknowledgements

This book took its time. I feel tremendously grateful to everyone who allowed it to do so, and who shaped it in many invaluable respects along the way. And I hereby forgive all those friends who thought it would be funny to point out that a book on the finitude of time was taking up so much of it. (And it *was* funny, sort of, on the first couple of occasions . . .)

The project would have got precisely nowhere without Tina Bennett, whom I thank both for her expert guidance and for many insights embedded in this book; later on, it benefited immeasurably from the involvement of Claire Conrad and Melissa Flashman at Janklow & Nesbit. I was also fortunate to work with Tracy Fisher at WME, along with her colleague in London, Matilda Forbes Watson. Among the numerous people at FSG to whom I owe gratitude, I'll mention here specifically my editor Eric Chinski, who (besides demonstrating enormous patience) vastly improved the text and pushed me to get much clearer on my ideas; and Julia Ringo, for handling the intricate later stages of editing with such expertise. I benefitted greatly from the skills of Lottchen Shivers and her colleagues in the publicity department, along with Daniel del Valle and (at Hilsinger-Mendelson) Sandi Mendelson, Emily Willette and Caroline Connors. Thanks also to Judy Kiviat, Maureen Klier, Christine Paik and Chris Peterson.

It has been a pleasure to work with Stuart Williams at The Bodley Head, who provided numerous important editorial comments reflected in the text, along with his exceedingly capable colleagues Aidan O'Neill, Sophie Painter and Mia Quibell-Smith; much thanks also to Beth Coates and her colleagues at Vintage. That so much time and attention on both sides of the Atlantic were bestowed while schools and offices were closed as a result of the coronavirus pandemic makes me all the more appreciative.

I first explored many of the topics discussed here in other venues, working with talented people including Melissa Denes, Paul Laity, Ruth Lewy, Jonathan Shainin, and David Wolf at the *Guardian*; Zan Boag at *New Philosopher*; and Peter McManus at the BBC. Conversations with Rachael Burnett, Lila Cecil, Jon Krop, Robin Parmiter, and Rachel Sherman were crucial to these ideas taking shape as a book. The following people also generously contributed their wisdom in the course of my research: Jessica Abel, Jim Benson, Stephanie Brown, Carl Cederström, James Hollis, Derrick Jensen, the late Robert Levine, Geoff Lye, María Martinón-Torres, Jennifer Roberts, Derek Sivers, Michael Taft, Antina von Schnitzler, Rebecca Wragg Sykes and Shinzen Young.

Ashley Tuttle provided a wonderful place to work at a critical moment, and I was lucky, once again, to write much of the rest of this book at Brooklyn Creative League, where Neil Carlson and Erin Carney have established a warm and supportive community. I'm very thankful for the friendship and conversation of Kenneth Folk and Maxson McDowell as well.

One temporal threshold I crossed while writing this book: I've now known Emma Brockes for more of my life than I haven't. I'm very glad about that, and for the fact that our children are now friends. Many conversations with her, some of which involved her talking me off metaphorical ledges, went into this book. Thank you also to my parents, Steven Burkeman and Jane Gibbins; my friends from York; my sister, Hannah, along with Alton, Layla and Ethan; Jeremy, Julia and Mari; Merope Mills; June Chaplin; and the Crawford-Montandon family.

No handful of sentences will do justice to Heather Chaplin's role in my life, but let me say here anyway how ridiculously grateful I am for her love, partnership, humour and integrity, and for the many sacrifices she made for this book. Our son, Rowan, arrived not long after work on it began. It would be a mischaracterisation (let's put it that way) to suggest that this development helped speed the book towards completion, but the transformative experience of getting to know him is certainly reflected in these pages. Unlimited love to you both.

My dear grandmother Erica Burkeman, whose childhood departure from Nazi Germany I describe in chapter 7, died in 2020 at the age of ninety-six. I don't know whether she would have read this book, but she would definitely have told everyone she met that I had written it.

Index

INDEX

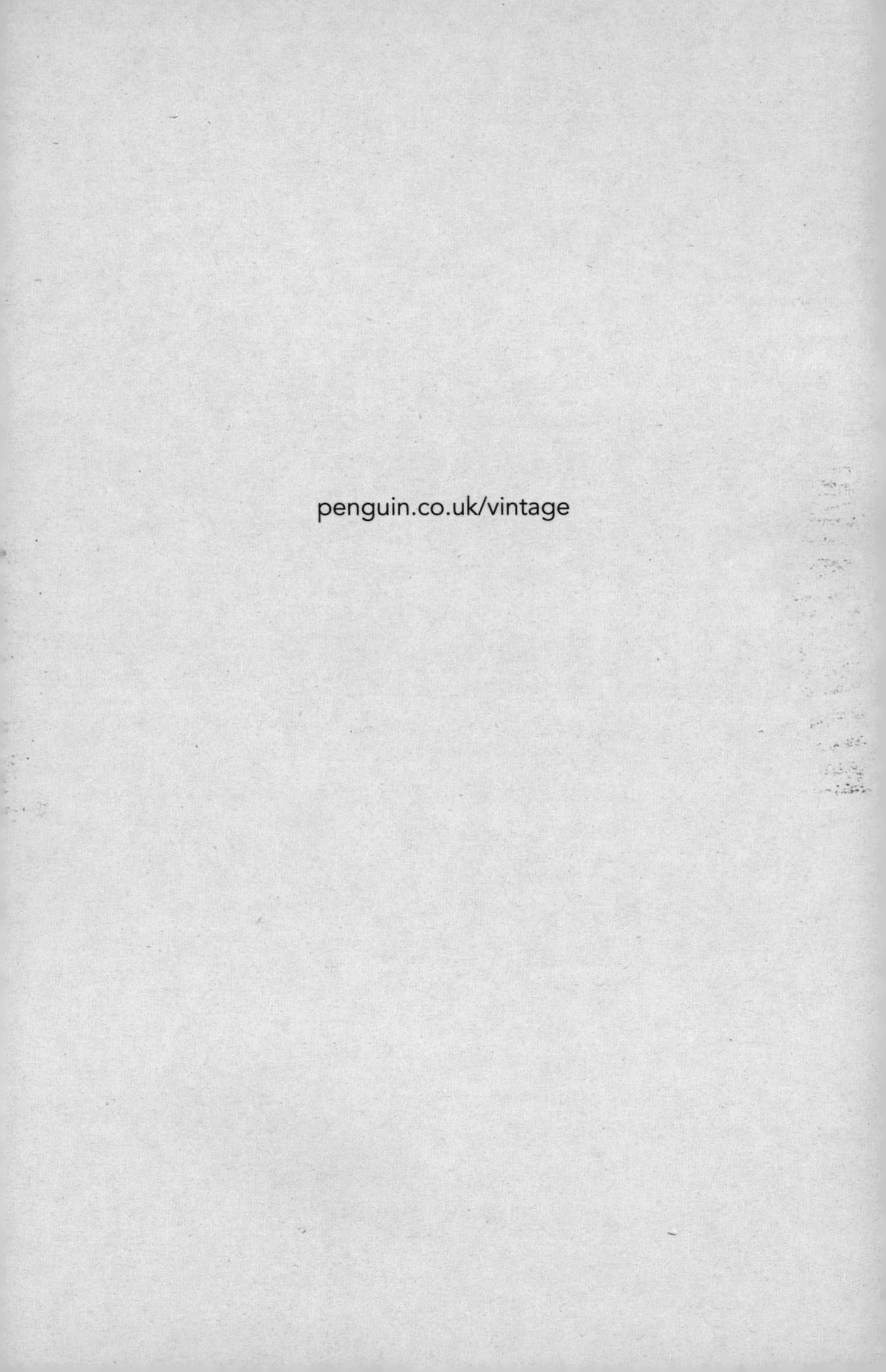
penguin.co.uk/vintage